Selected Papers from the "7th Workshop on the Nuclear Mass Table with DRHBc Theory"

Selected Papers from the "7th Workshop on the Nuclear Mass Table with DRHBc Theory"

Guest Editors

Shuangquan Zhang
Youngman Kim

MDPI

Basel • Beijing • Wuhan • Barcelona • Belgrade • Novi Sad • Cluj • Manchester

Guest Editors
Shuangquan Zhang
School of Physics
Peking University
Beijing
China

Youngman Kim
Center for Exotic Nuclear Studies
Institute for Basic Science
Daejeon
South Korea

Editorial Office
MDPI AG
Grosspeteranlage 5
4052 Basel, Switzerland

This is a reprint of the Special Issue, published open access by the journal *Particles* (ISSN 2571-712X), freely accessible at: https://www.mdpi.com/journal/particles/special_issues/DRHBc2024.

For citation purposes, cite each article independently as indicated on the article page online and as indicated below:

Lastname, A.A.; Lastname, B.B. Article Title. *Journal Name* **Year**, *Volume Number*, Page Range.

ISBN 978-3-7258-4411-1 (Hbk)
ISBN 978-3-7258-4412-8 (PDF)
https://doi.org/10.3390/books978-3-7258-4412-8

Contents

About the Editors

Shuangquan Zhang

Shuangquan Zhang received his Ph.D. in Particle Physics and Nuclear Physics from Peking University in 2002. His research primarily focuses on nuclear structure and nuclear astrophysics. He has conducted systematic investigations in several key areas, including nuclear chiral symmetry, novel collective excitation modes, the properties of exotic nuclei, and relativistic density functional theory. He currently serves as an Associate Professor at Peking University.

Youngman Kim

Youngman Kim earned his Ph.D. in Hadron and Nuclear Physics from Hanyang University in 1999.

His current research interests focus on RAON-related physics, including nuclear reactions, nuclear structure, and nuclear astrophysics.

As the leader of the nuclear theory group at the Institute for Basic Science (IBS), he has successfully established a RAON-focused nuclear theory team within the institute.

Preface

In December 2018, a campaign was launched to construct the first nuclear mass table that simultaneously incorporates continuum and deformation effects. This ambitious initiative emerged from the increasing demand to acquire mass data of experimentally-challenging exotic nuclei and to deepen our understanding of nuclear physics, leveraging the remarkable capabilities of the deformed relativistic Hartree–Bogoliubov theory in continuum (DRHBc). To achieve this goal, an international collaboration—the DRHBc mass table collaboration—was established. The collaboration now comprises 33 universities and institutions from countries such as China, South Korea, and Japan.

Over the past seven years, the DRHBc mass table collaboration has achieved several significant milestones. By integrating the cutting-edge DRHBc with the superior density functional PC-PK1, the collaboration has produced two widely cited compilations in *Atomic Data and Nuclear Data Tables*. These publications summarizes the ground-state properties of even–even and even–Z nuclei with $8 \leq Z \leq 120$, covering the region from the proton drip line to the neutron drip line. Beyond constructing the nuclear mass table, the collaboration has also conducted a series of remarkable studies on a wide range of topics, including deformed halos, charge radii, stability peninsula, structural evolution, prolate dominance, shape coexistence, dynamical correlations, and particle emissions, among others.

A series of workshops have played a crucial role in facilitating this collaborative effort. The inaugural workshop, held in Daejeon and Seoul, laid the foundation for the formation of a vibrant and productive research community. Subsequent workshops, hosted at various locations across Asia, have marked the continued growth of this community, attracting an increasing number of participants from diverse institutions and fostering new research networks. The "7th workshop on nuclear mass table with DRHBc theory", hosted by Soongsil University and held in Gangneung from July 1 to 4, 2024, represents a milestone in this ongoing journey. This workshop focused on the final stage of constructing the DRHBc mass table, with a particular emphasis on the properties of odd-Z nuclei and superheavy nuclei. Looking ahead, we remain committed to pursuing excellence in this field and eagerly anticipate the new discoveries that lie ahead.

This Special Issue features 10 contributions from esteemed experts who participated in the 7th workshop. It aims to provide a comprehensive overview of the current status of the DRHBc mass table project and explore related key issues, including structural evolution in various nuclear regions, possible magic numbers in superheavy nuclei, and a range of exotic nuclear phenomena. While this Special Issue is primarily intended for the nuclear physics community, this collection may also be of interest to other branches of physics.

In closing, we would like to take this opportunity to express our deep gratitude to Prof. Jie Meng, the Chair of the workshop series, for proposing this Special Issue, and to all the contributing authors for their valuable work. We also sincerely acknowledge the efforts of the MDPI Book staff and the *Particles* Editorial team. Special thanks are extended to Mr. Ethan Zhang, Managing Editor, for his dedicated assistance throughout the preparation of this volume.

Shuangquan Zhang and Youngman Kim
Guest Editors

particles

MDPI

Article

Shape Coexistence in Odd-Z Isotopes from Fluorine to Potassium

Myeong-Hwan Mun [1,2], Panagiota Papakonstantinou [3] and Youngman Kim [4,*]

1 Department of Physics, Kyungpook National University, Daegu 41566, Republic of Korea; aa3101@gmail.com
2 Department of Physics and Origin of Matter and Evolution of Galaxies (OMEG) Institute, Soongsil University, Seoul 06978, Republic of Korea
3 Institute for Rare Isotope Science, Institute for Basic Science, Daejeon 34000, Republic of Korea; ppapakon@ibs.re.kr
4 Center for Exotic Nuclear Studies, Institute for Basic Science, Daejeon 34126, Republic of Korea
* Correspondence: ykim@ibs.re.kr

Abstract: The shape of a nucleus is one of its fundamental properties. We conduct a systematic investigation of shape coexistence in odd-Z nuclei from fluorine to potassium using the deformed relativistic Hartree–Bogoliubov theory in continuum. First, we present a simple argument regarding the energy differences between degenerate vacua, which can serve as a criterion for identifying candidates for shape coexistence. We then predict isotopes that exhibit shape coexistence.

Keywords: relativistic density functional theory; shape coexistence; potential energy curve

1. Introduction

A nucleus, composed of nucleons that can be thought of as quantum marbles, is a fascinating entity exhibiting a variety of characteristics due to the quantum nature of its constituents. Understanding how nucleons combine to form a nucleus with exotic features is a crucial problem that has garnered significant attention in recent years, especially with the advent of new rare isotope beam facilities that enable the production of more exotic nuclei. Ultimately, we must explore how the shell and collective properties of a nucleus emerge from fundamental theories such as quantum chromodynamics. The shape of a nucleus is one of its most fundamental properties; some nuclei display exotic characteristics such as pear shapes, bubble structures, and shape coexistence. The investigation of these exotic nuclear features provides invaluable insights into the complex nature of nuclear forces that govern the formation and structure of atomic nuclei. By exploring how neutrons and protons arrange themselves within these unusual shapes, one can gain a deeper understanding of the underlying nuclear forces and quantum mechanical effects at play.

Shape coexistence is a noteworthy exotic property of nuclei [1–6]. A nucleus can display different shapes with only a small energy difference compared to its total binding energy. Nuclei that exhibit shape coexistence have multiple minima in their potential energy curve. Shape coexistence is closely linked to the island of inversion [7,8] because the change in the ordering of nuclear shells in an island of inversion can result in the emergence of multiple nuclear shapes. Islands of inversion have been studied by a variety of methods, e.g., large-scale shell model calculations [9] and ab initio calculations [10].

In density functional theory, shape coexistence is closely linked to the existence of degenerate vacua. This raises the following question: how small is "small" in the context of degenerate vacua? We can roughly estimate this small energy difference using the uncertainty principle. Taking $\Delta x \cdot \Delta p \approx \hbar/2$ with a nuclear diameter $\Delta x \approx 2.5A^{1/3}$ fm, we can estimate the energy uncertainty as $\Delta E \approx (\Delta p)^2/2m \approx 3.3/A^{2/3}$ MeV $\approx$ 100–300 keV

 https://doi.org/10.3390/particles8010032

for most nuclei. Therefore, in the context of density functional theory, we can consider an energy difference of a few hundred keV to be "small".

The main motivation for the present theoretical survey of potential candidates for phase coexistence is the possibility to study the phenomenon in stable and exotic light nuclei in new RI facilities, and especially RAON in South Korea [11]. In a previous work [12], we focused on even$-Z$ candidates in the region from oxygen to calcium. At present, we expand our study to odd$-Z$ isotopes in the same region, from fluorine (F) to potassium (K). As in that earlier study, we use the deformed relativistic Hartree–Bogoliubov theory in continuum (DRHBc) and the PC-PK1 density functional [13]. This approach allows us to investigate the properties of both exotic and stable nuclei by self-consistently incorporating axial deformation, pairing correlations, and continuum effects [14–20]. We refer to ref. [21] for a review of shape coexistence in odd-mass nuclei and refs. [22,23] for some recent studies on shape coexistence.

2. Deformed Relativistic Hartree–Bogoliubov Theory in Continuum

The Lagrangian density of DRHBc is given by [13,19,24]:

$$\begin{aligned}
\mathcal{L} \;=\;& \bar{\psi}(i\gamma_\mu\partial^\mu - m)\psi - \frac{1}{2}\alpha_S(\bar{\psi}\psi)(\bar{\psi}\psi) - \frac{1}{2}\alpha_V(\bar{\psi}\gamma_\mu\psi)(\bar{\psi}\gamma^\mu\psi) \\
&-\frac{1}{2}\alpha_{TV}(\bar{\psi}\vec{\tau}\gamma_\mu\psi)(\bar{\psi}\vec{\tau}\gamma^\mu\psi) - \frac{1}{3}\beta_S(\bar{\psi}\psi)^3 - \frac{1}{4}\gamma_S(\bar{\psi}\psi)^4 - \frac{1}{4}\gamma_V\left[(\bar{\psi}\gamma_\mu\psi)(\bar{\psi}\gamma^\mu\psi)\right]^2 \\
&-\frac{1}{2}\delta_S\partial_\nu(\bar{\psi}\psi)\partial^\nu(\bar{\psi}\psi) - \frac{1}{2}\delta_V\partial_\nu(\bar{\psi}\gamma_\mu\psi)\partial^\nu(\bar{\psi}\gamma^\mu\psi) \\
&-\frac{1}{2}\delta_{TV}\partial_\nu(\bar{\psi}\vec{\tau}\gamma_\mu\psi)\partial^\nu(\bar{\psi}\vec{\tau}\gamma_\mu\psi) - \frac{1}{4}F^{\mu\nu}F_{\mu\nu} - e\frac{1-\tau_3}{2}\bar{\psi}\gamma^\mu\psi A_\mu,
\end{aligned} \tag{1}$$

where m represents the nucleon mass and α_S, α_V, and α_{TV} denote the coupling constants for four-fermion contact interactions. The terms involving β_S, γ_S, and γ_V account for density-dependent effects, while those with δ_S, δ_V, and δ_{TV} reflect finite-range effects. Additionally, A_μ and $F_{\mu\nu}$ correspond to the four-vector potential and the electromagnetic field strength tensor, respectively. The subscripts S, V, and TV stand for scalar, vector, and isovector, respectively.

By applying the mean-field approximation to the Lagrangian density in Equation (1) and performing the Legendre transformation, we derive the mean-field Hamiltonian density. Using the variational method on this Hamiltonian density, we then arrive at the relativistic Hartree–Bogoliubov equation [25].

$$\begin{pmatrix} h_D - \lambda & \Delta \\ -\Delta^* & -h_D^* + \lambda \end{pmatrix}\begin{pmatrix} U_k \\ V_k \end{pmatrix} = E_k\begin{pmatrix} U_k \\ V_k \end{pmatrix}. \tag{2}$$

Here, E_k denotes the quasiparticle energy, and U_k and V_k are the quasiparticle wave functions, with λ denoting the Fermi energy. The Dirac Hamiltonian h_D is given by

$$h_D(\boldsymbol{r}) = \boldsymbol{\alpha}\cdot\boldsymbol{p} + \beta(M + S(\boldsymbol{r})) + V(\boldsymbol{r}), \tag{3}$$

and the scalar $S(\boldsymbol{r})$ and vector $V(\boldsymbol{r})$ potentials can be expressed as

$$S(\boldsymbol{r}) = \alpha_S\rho_S + \beta_S\rho_S^2 + \gamma_S\rho_S^3 + \delta_S\Delta\rho_S, \tag{4}$$

$$\begin{aligned}
V(\boldsymbol{r}) =\;& \alpha_V\rho_V + \gamma_V\rho_V^3 + \delta_V\Delta\rho_V + eA_0 \\
&+ \alpha_{TV}\tau_3\rho_{TV} + \delta_{TV}\tau_3\Delta\rho_{TV}.
\end{aligned} \tag{5}$$

The local densities $\rho_S(\boldsymbol{r})$, $\rho_V(\boldsymbol{r})$, and $\rho_{TV}(\boldsymbol{r})$ can be expressed in terms of the quasiparticle wave functions as follows:

$$\rho_S(\boldsymbol{r}) = \sum_{k>0} \bar{V}_k(\boldsymbol{r}) V_k(\boldsymbol{r}), \tag{6}$$

$$\rho_V(\boldsymbol{r}) = \sum_{k>0} V_k^{\dagger}(\boldsymbol{r}) V_k(\boldsymbol{r}), \tag{7}$$

$$\rho_{TV}(\boldsymbol{r}) = \sum_{k>0} V_k^{\dagger}(\boldsymbol{r}) \tau_3 V_k(\boldsymbol{r}). \tag{8}$$

In principle, we can derive the pairing potential for the particle–particle channel from the Lagrangian density in Equation (1), but for simplicity, we adopt the following form:

$$\Delta_{kk'}(\boldsymbol{r}, \boldsymbol{r}') = -\sum_{\tilde{k}\tilde{k}'} V^{pp}_{kk',\tilde{k}\tilde{k}'}(\boldsymbol{r}, \boldsymbol{r}') \kappa_{\tilde{k}\tilde{k}'}(\boldsymbol{r}, \boldsymbol{r}'), \tag{9}$$

where the pairing tensor is defined by $\kappa = V^* U^T$. For V^{pp}, we use the density-dependent zero-range pairing interaction

$$V^{pp}(\boldsymbol{r}, \boldsymbol{r}') = \frac{V_0}{2}(1 - P^{\sigma}) \delta(\boldsymbol{r} - \boldsymbol{r}') \left(1 - \frac{\rho(\boldsymbol{r})}{\rho_{sat}}\right), \tag{10}$$

where ρ_{sat} is the nuclear saturation density. The total energy of a nucleus can be expressed as

$$\begin{aligned} E_{\text{tot}} &= \sum_{k>0} (\lambda_\tau - E_k) v_k^2 - E_{\text{pair}} + E_{\text{c.m.}} - \int d^3\mathbf{r} \left[\frac{1}{2}\alpha_S \rho_S^2 \right. \\ &+ \frac{1}{2}\alpha_V \rho_V^2 + \frac{1}{2}\alpha_{TV}\rho_{TV}^2 + \frac{2}{3}\beta_S \rho_S^3 + \frac{3}{4}\gamma_S \rho_S^4 + \frac{3}{4}\gamma_V \rho_V^4 \\ &+ \left. \frac{1}{2}(\delta_S \rho_S \Delta\rho_S + \delta_V \rho_V \Delta\rho_V + \delta_{TV}\rho_3 \Delta\rho_3 + \rho_p e A^0) \right], \end{aligned} \tag{11}$$

where $E_{\text{c.m.}}$ denotes the center of mass correction energy. The zero-range pairing force results in a local pairing field $\Delta(\boldsymbol{r})$ with the associated pairing energy energy expressed as follows:

$$E_{\text{pair}} = -\frac{1}{2} \int d^3\boldsymbol{r} \kappa(\boldsymbol{r}) \Delta(\boldsymbol{r}). \tag{12}$$

To investigate exotic nuclear properties, it is essential to self-consistently incorporate both continuum and deformation effects, as well as pairing. We expand the wave functions using the Dirac Wood–Saxon basis to account for continuum effects. To address axial deformation while maintaining spatial reflection symmetry, we expand the potentials ($S(\boldsymbol{r})$ and $V(\boldsymbol{r})$) and densities ($\rho_S(\boldsymbol{r})$, $\rho_V(\boldsymbol{r})$, and $\rho_{TV}(\boldsymbol{r})$) in terms of Legendre polynomials. For more details on DRHBc, we refer to [17,19].

By performing constrained calculations for different degrees of quadrupole deformation, we can obtain not only the lowest energy solution for a given nucleus, but also identify other possible local minima in the potential energy curves. The existence of two near-degenerate vacua will lead to two perturbative ground states 1 and 2. Neither of them is the exact ground state. The ground state is given by the symmetric or the antisymmetric linear combination of the perturbative ground states.

3. Results

In this section, we present the candidate isotopes exhibiting shape coexistence for fluorine (F, $Z = 9$), sodium (Na, $Z = 11$), aluminium (Al, $Z = 13$), phosphorus (P, $Z = 15$), chlorine (Cl, $Z = 17$), and potassium (K, $Z = 19$).

To identify candidate isotopes that exhibit shape coexistence, we examine the potential energy curve (PEC) of each isotope as a function of the quadrupole deformation parameter (β_2). We search for isotopes with near-degenerate minima. Examples of PECs for potential candidates, namely ^{23}F, ^{31}Al, ^{38}Cl, and ^{42}K, are shown in Figure 1. The PECs are compared with existing results obtained with the Hartree–Fock–Bogoliubov approach using the Gogny D1S functional, which are available at [26] and found qualitatively similar.

Based on our rough estimates using the uncertainty principle outlined in the Introduction, we find that the energy difference ΔE for shape coexistence is typically on the order of a few hundreds keV, as also shown in ref. [27]. In this work, since we are studying light nuclei, which generally have lower total energies compared to medium- or heavy-mass nuclei, we adopt the criterion of 300 keV for shape coexistence.

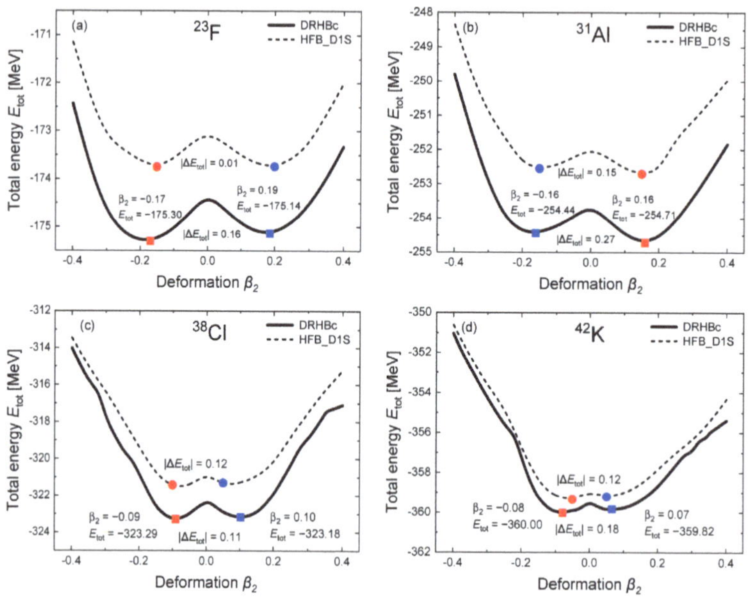

Figure 1. Potential energy curves for selected isotopes (**a**) ^{23}F, (**b**) ^{31}Al, (**c**) ^{38}Cl, and (**d**) ^{42}K as a function of the quadrupole deformation parameter (β_2). Results of the present DRHBc calculations (solid line) are shown in comparison with those of HFB calculations (dash line) with the Gogny D1S effective interaction [26]. The square and the circle represent local minima of DRHBc and HFB calculations, respectively.

In Table 1, we list all forty-six odd-Z isotopes in the range $Z = 9$ to $Z = 19$ that are found to have two minima with an energy difference ΔE smaller than 300 keV. Expanding the ΔE criterion to below 500 keV leads to twenty-four additional candidates for shape coexistence, which are not listed in Table 1, but are indicated in Figure 2.

Figure 2 shows all odd-Z nuclei from F to K which are predicted particle-bound using the DRHBc theory with the PC-PK1 functional. The candidates for shape coexistence in odd-Z isotopes are highlighted in red text within the yellow boxes ($\Delta E \leq 300$ keV) and cyan boxes ($\Delta E \leq 500$ keV). In addition, the candidates based on results with the Gogny D1S functional, available at [26], are shown with red circles. The regions of candidates predicted with the two approaches and functionals are similar, which is not surprising given that they both are mean-field approaches. Experiments on nuclei where there are deviations could offer valuable insights for further theoretical developments.

Table 1. The odd-Z isotopes we found in the range Z = 9 to Z = 19 that have two minima with an energy difference ΔE smaller than 300 keV.

Isotopes	β_2	ΔE (MeV)	Isotopes	β_2	ΔE (MeV)
^{16}F	−0.12, 0.20	0.09	^{31}Cl	−0.17, 0.09	0.27
^{17}F	−0.08, 0.10	0.03	^{32}Cl	−0.10, 0.08	0.12
^{23}F	−0.17, 0.19	0.16	^{33}Cl	−0.07, 0.06	0.08
^{24}F	−0.13, 0.13	0.13	^{38}Cl	−0.09, 0.10	0.11
^{25}F	−0.08, 0.09	0.04	^{39}Cl	−0.09, 0.12	0.05
^{26}F	−0.12, 0.13	0.02	^{44}Cl	−0.22, 0.27	0.17
^{27}F	−0.08, 0.10	0.02	^{50}Cl	−0.25, 0.06	0.23
^{28}F	−0.09, 0.09	0.02	^{51}Cl	−0.21, 0.10	0.21
^{25}Na	−0.26, 0.31	0.12	^{55}Cl	−0.10, 0.08	0.27
^{30}Na	−0.08, 0.10	0.10	^{56}Cl	−0.06, 0.06	0.06
^{27}Al	−0.27, 0.18	0.18	^{33}K	−0.09, 0.06	0.12
^{28}Al	−0.19, 0.16	0.21	^{34}K	−0.06, 0.05	0.07
^{31}Al	−0.16, 0.16	0.27	^{35}K	−0.05, 0.05	0.04
^{32}Al	−0.11, 0.11	0.25	^{37}K	−0.08, 0.04	0.21
^{33}Al	−0.04, 0.06	0.12	^{38}K	−0.07, 0.05	0.23
^{41}Al	−0.36, 0.34	0.24	^{40}K	−0.07, 0.06	0.23
^{45}Al	−0.26, 0.19	0.29	^{41}K	−0.06, 0.04	0.09
^{28}P	−0.20, 0.16	0.20	^{42}K	−0.08, 0.07	0.18
^{32}P	−0.10, 0.08	0.15	^{43}K	−0.10, 0.06	0.28
^{34}P	−0.06, 0.05	0.08	^{50}K	−0.10, 0.07	0.20
^{36}P	−0.07, 0.10	0.02	^{51}K	−0.08, 0.04	0.10
^{47}P	−0.25, 0.13	0.13	^{52}K	−0.05, 0.03	0.03
^{50}P	−0.05, 0.08	0.07	^{53}K	−0.07, 0.03	0.20

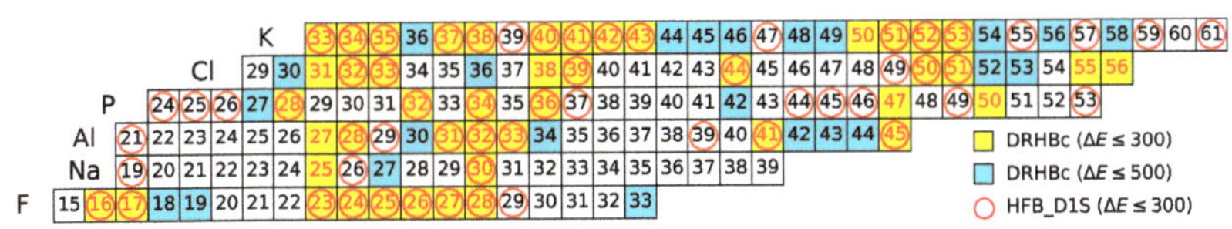

Figure 2. Summary of nuclei exhibiting shape coexistence in odd-Z isotopes from F to K. The shape coexistence candidates are highlighted in red text within yellow boxes ($\Delta E \leq 300$ keV). The x-axis represents the mass number of the isotopes. The cyan boxes represent 24 additional candidates based on the 500 keV criterion. Only odd-Z isotopes are shown for clarity.

It is worth noting that the dual-shell mechanism, used to predict regions of the nuclear chart where shape coexistence might occur, was introduced in refs. [28,29]. The studies showed that nuclei with proton or neutron numbers in the ranges 7–8, 17–20, 34–40, 59–70, 96–112, and 146–168 are potential candidates for shape coexistence. Let us take Cl isotopes as an example to compare our results with the prediction in refs. [28,29]. In our study, using the 500 MeV criterion, we identify 14 candidate isotopes, 7 of which align with the results from refs. [28,29].

4. Summary

We identified odd-Z isotopes from fluorine to potassium which are predicted to show shape coexistence within the DRHBc framework. Based on the uncertainty principle, we estimated that the energy difference ΔE between degenerate minima on the potential energy curve for shape coexistence has to be of the order of a few hundred keV. In this work, we adopted a criterion of 300 keV (500 keV) for shape coexistence in light nuclei and identified 46 (70) candidate isotopes from fluorine to potassium.

Author Contributions: Calculations—formal analysis, M.-H.M.; writing—original draft preparation, Y.K.; review and editing, P.P. All authors have read and agreed to the published version of the manuscript.

Funding: This work was supported in part by the Institute for Basic Science (IBS-R031-D1, 2013M7A1A1075764). M.-H.M. was supported by the National Research Foundation of Korea NRF grants funded by the Korean government Ministry of Science and ICT (grant no. NRF-2021R1F1A1060066).

Informed Consent Statement: Not applicable.

Data Availability Statement: The dataset can be accessed upon request to the corresponding author.

Acknowledgments: The authors would like to express their gratitude to the members of the DRHBc Mass Table Collaboration for their valuable discussions.

Conflicts of Interest: The authors declare no conflicts of interest.

References

1. Wood, J.L.; Heyde, K.; Nazarewicz, W.; Huyse, M.; Van Duppen, P. Coexistence in even-mass nuclei. *Phys. Rep.* **1992**, *215*, 101–201. [CrossRef]
2. Gade, A.; Liddick, S. Shape coexistence in neutron-rich nuclei. *J. Phys. G* **2016**, *43*, 024001. [CrossRef]
3. Garrett, P.E.; Zielińska, M.; Clément, E. An experimental view on shape coexistence in nuclei. *Prog. Part. Nucl. Phys.* **2022**, *124*, 103931. [CrossRef]
4. Choi, Y.B.; Lee, C.H.; Mun, M.H.; Kim, Y. Bubble nuclei with shape coexistence in even-even isotopes of Hf to Hg. *Phys. Rev. C* **2022**, *105*, 024306. [CrossRef]
5. Bonatsos, D.; Martinou, A.; Peroulis, S.K.; Mertzimekis, T.J.; Minkov, N. Shape Coexistence in Even–Even Nuclei: A Theoretical Overview. *Atoms* **2023**, *11*, 117. [CrossRef]
6. Leoni, S.; Fornal, B.; Bracco, A.; Tsunoda, Y.; Otsuka, T. Multifaceted character of shape coexistence phenomena in atomic nuclei. *Prog. Part. Nucl. Phys.* **2024**, *139*, 104119. [CrossRef]
7. Brown, B.A. Islands of insight in the nuclear chart. *Physics* **2010**, *3*, 104. [CrossRef]
8. Scheit, H. Spectroscopy in and around the Island of inversion. *J. Phys. Conf. Ser.* **2011**, *312*, 092010. [CrossRef]
9. Caurier, E.; Nowacki, F.; Poves, A. Merging of the islands of inversion at $N = 20$ and $N = 28$. *Phys. Rev. C* **2014**, *90*, 014302. [CrossRef]
10. Miyagi, T.; Stroberg, S.R.; Holt, J.D.; Shimizu, N. Ab initio multishell valence-space Hamiltonians and the island of inversion. *Phys. Rev. C* **2020**, *102*, 034320. [CrossRef]
11. Jeong, S.; Papakonstantinou, P.; Ishiyama, H.; Kim, Y. A brief overview of RAON physics. *JKPS* **2018**, *73*, 516. [CrossRef]
12. In, E.J.; Kim, Y.; Papakonstantinou, P.; Hong, S.W. Shape coexistence in isotopes from oxygen to calcium. *JKPS* **2020**, *77*, 966. [CrossRef]
13. Zhao, P.W.; Li, Z.P.; Yao, J.M.; Meng, J. New parametrization for the nuclear covariant energy density functional with a point-coupling interaction. *Phys. Rev. C* **2010**, *82*, 054319. [CrossRef]
14. Zhou, S.G.; Meng, J.; Ring, P.; Zhao, E.G. Neutron halo in deformed nuclei. *Phys. Rev. C* **2010**, *82*, 011301. [CrossRef]
15. Li, L.; Meng, J.; Ring, P.; Zhao, E.G.; Zhou, S.G. Deformed relativistic Hartree-Bogoliubov theory in continuum. *Phys. Rev. C* **2012**, *85*, 024312. [CrossRef]
16. Meng, J.; Zhou, S.G. Halos in medium-heavy and heavy nuclei with covariant density functional theory in continuum. *J. Phys. G* **2015**, *42*, 093101. [CrossRef]
17. Zhang, K.; Cheoun, M.K.; Choi, Y.B.; Chong, P.S.; Dong, J.; Geng, L.; Ha, E.; He, X.; Heo, C.; Ho, M.C.; et al. Deformed relativistic Hartree-Bogoliubov theory in continuum with a point-coupling functional: Examples of even-even Nd isotopes. *Phys. Rev. C* **2020**, *102*, 024314. [CrossRef]
18. Zhang, K.; Cheoun, M.K.; Choi, Y.B.; Chong, P.S.; Dong, J.; Dong, Z.; Du, X.; Geng, L.; Ha, E.; He, X.T.; et al. Nuclear mass table in deformed relativistic Hartree–Bogoliubov theory in continuum, I: Even–even nuclei. *At. Data Nucl. Data Tables* **2022**, *144*, 101488. [CrossRef]
19. Pan, C.; Cheoun, M.K.; Choi, Y.B.; Dong, J.; Du, X.; Fan, X.H.; Gao, W.; Geng, L.; Ha, E.; He, X.T.; et al. Deformed relativistic Hartree-Bogoliubov theory in continuum with a point-coupling functional. II. Examples of odd Nd isotopes. *Phys. Rev. C* **2022**, *106*, 014316. [CrossRef]

20. Guo, P.; Cao, X.; Chen, K.; Chen, Z.; Cheoun, M.K.; Choi, Y.B.; Lam, P.C.; Deng, W.; Dong, J.; Du, P.; et al. Nuclear mass table in deformed relativistic Hartree–Bogoliubov theory in continuum, II: Even-Z nuclei. *At. Data Nucl. Data Tables* **2024**, *158*, 101661. [CrossRef]
21. Heyde, K.; Van Isacker, P.; Waroquier, M.; Wood, J.L.; Meyer, R.A. Coexistence in odd-mass nuclei. *Phys. Rep.* **1983**, *102*, 291. [CrossRef]
22. Aggarwal, M. Shape coexistence in excited odd-Z proton emitters Eu131−136. *Phys. Rev. C* **2014**, *90*, 064322. [CrossRef]
23. Sun, Q.K.; Sun, T.T.; Zhang, W.; Zhang, S.S.; Chen, C. Possible shape coexistence in odd-A Ne isotopes and the impurity effects of Λ hyperons. *Chin. Phys. C* **2022**, *46*, 074106. [CrossRef]
24. Meng, J. *Relativistic Density Functional for Nuclear Structure*; World Scientific: Singapore, 2016.
25. Kucharek, H.; Ring, P.; Schuk, P.; Bengtsson, R.; Girod, M. Pairing properties of nuclear matter from the gogny force. *Phys. Lett. B* **1989**, *216*, 249. [CrossRef]
26. Available online: https://www-phynu.cea.fr/science_en_ligne/carte_potentiels_microscopiques/carte_potentiel_nucleaire_eng.htm (accessed on 4 February 2025).
27. Abusara, H.; Ahmad, S. Shape evolution in Kr, Zr, and Sr isotopic chains in covariant density functional theory. *Phys. Rev. C* **2017**, *96*, 064303. [CrossRef]
28. Martinou, A.; Bonatsos, D.; Mertzimekis, T.J.; Karakatsanis, K.E.; Assimakis, I.E.; Peroulis, S.K.; Sarantopoulou, S.; Minkov, N. The islands of shape coexistence within the Elliott and the proxy-SU(3) Models. *Eur. Phys. J. A* **2021**, *57*, 84. [CrossRef]
29. Martinou, A.; Bonatsos, D.; Peroulis, S.K.; Karakatsanis, K.E.; Mertzimekis, T.J.; Minkov, N. Islands of Shape Coexistence: Theoretical Predictions and Experimental Evidence. *Symmetry* **2023**, *15*, 29. [CrossRef]

particles

MDPI

Article

Giant Halo in ^{66}Ca Within Relativistic Continuum Hartree–Bogoliubov Theory Combined with Lipkin–Nogami Method

Chang Zhou, Peng Guo and Xiaofei Jiang *

State Key Laboratory of Nuclear Physics and Technology, School of Physics, Peking University, Beijing 100871, China; c.zhou@pku.edu.cn (C.Z.); 2301110125@pku.edu.cn (P.G.)
* Correspondence: jiangxiaofei@pku.edu.cn

Abstract: The impact of the Lipkin–Nogami (LN) method on a giant halo is investigated within the relativistic continuum Hartree–Bogoliubov (RCHB) theory. The ground-state properties of Ca isotopes obtained from RCHB and RCHB+LN calculations are presented. The results show that the LN correction does not change the range of Ca isotopes with a giant halo. Taking ^{66}Ca as an example, the neutron density distribution with LN correction is found to be slightly more diffused, which can be illustrated by the enlargement of the root mean square radius and the enhancement of the relative contribution in neutron $3s_{1/2}$ level.

Keywords: relativistic continuum Hartree–Bogoliubov theory; Lipkin–Nogami method; Dirac Woods–Saxon basis; giant halo

1. Introduction

The study of exotic nuclei with extreme N/Z ratios has become one of the frontiers of nuclear physics [1–4]. One of the interesting and impressive phenomena in exotic nuclei is the halo, which was first discovered in ^{11}Li [5] and has been observed in many light nuclei near drip lines [6]. Experimentally, some unusual properties have been found in halo nuclei, for example, their abnormally large radii [5], their narrow momentum distribution in the breakup reaction [7,8], the enhancement of electromagnetic dissociation cross-sections [9], etc.

In halo nuclei, the weakly-bound valence nucleons could be easily scattered to the continuum states due to the pairing correlations [10]. Thus, a theory which gives a proper theoretical description of halo phenomena must properly treat the continuum, the pairing correlations, the large spatial distributions, and the coupling among them. The relativistic continuum Hartree–Bogoliubov (RCHB) theory [11] fulfills all these features self-consistently. It has achieved great success in reproducing and interpreting the halo in ^{11}Li [10]. The RCHB theory also predicted the halo nuclei in Zr isotopes with a large amount of particle numbers in the halo region [12]. This special kind of halo is called the giant halo. Giant halo phenomena have also been predicted in Ca isotopes [13,14] within RCHB theory. In addition to RCHB theory, the relativistic mean field–BCS approach combined with proper treatments of the resonant continuum [15–18], the relativistic [19] and non-relativistic [20–22] Hartree–Fock–Bogoliubov theory, and the RCHB theory with the Green's function method [23,24] also give the predictions of giant halos.

In RCHB theory, the pairing correlations are treated with Bogoliubov transformation. However, Bogoliubov transformation breaks the symmetry of the particle number. Thus, it is necessary to investigate the influence of the particle number symmetry breaking on the halo and giant halo. Variation after particle number projection method is a rigorous way to restore the broken symmetry of the particle number [25,26]. However, such calculations are expected to have a massive computational cost for realistic interactions [26,27]. To reduce the time consumption, an approximation to the variation after the particle number projection method by means of the Lipkin–Nogami (LN) method [28–30] has been widely used. Up to now, the LN method has been applied in many nuclear studies, such as nuclear charge distributions [31], abnormal odd–even staggering around ^{132}Sn [32], the properties of low-lying states [33,34], superdeformed bands [27,35,36], the thermodynamic properties of nuclei [37], and two-neutrino double beta decay [38].

In this paper, we combine the RCHB theory with the LN method to investigate the giant halo phenomena in Ca isotopes, especially in ^{66}Ca. By comparing the RCHB and RCHB+LN calculations, the impact of the LN correction on giant halos is discussed.

2. Method

2.1. The Relativistic Continuum Hartree–Bogoliubov Theory

The detailed formalism of RCHB theory can be found in Refs. [11,39,40]. The RCHB theory, assuming the spherical symmetry, solves the relativistic Hartree–Bogoliubov (RHB) equations [41] in the coordinate space in order to deal with the large density distribution. Alternatively, one can use a basis with improved asymptotic behavior at large distances [42], for example, the spherical Dirac Woods–Saxon (DWS) basis used here. As shown in Ref. [42], an expansion in the DWS basis is fully equivalent to the calculations in coordinate space. In fact, the DWS basis has already been used in the deformed relativistic Hartree–Bogoliubov theory in continuum (DRHBc) [43], which has many applications so far [44–48]. Our examination shows that the results of RCHB solved in the DWS basis are the same as those of DRHBc with a spherical shape.

The RHB equations self-consistently treat the mean field and pairing correlations,

$$\begin{pmatrix} h_D - \lambda & \Delta \\ -\Delta^* & -h_D^* + \lambda \end{pmatrix} \begin{pmatrix} U_k \\ V_k \end{pmatrix} = E_k \begin{pmatrix} U_k \\ V_k \end{pmatrix}, \tag{1}$$

where λ is the Fermi surface, E_k is the quasiparticle energy, and $(U_k, V_k)^T$ are the quasiparticle wave functions. The Dirac Hamiltonian h_D in coordinate space reads as

$$h_D(\boldsymbol{r}) = \boldsymbol{\alpha} \cdot \boldsymbol{p} + V(\boldsymbol{r}) + \beta[M + S(\boldsymbol{r})], \tag{2}$$

with $S(\boldsymbol{r})$ and $V(\boldsymbol{r})$ being the scalar and the vector potential, respectively, and where

$$S(\boldsymbol{r}) = \alpha_S \rho_S + \beta_S \rho_S^2 + \gamma_S \rho_S^3 + \delta_S \Delta \rho_S, \tag{3}$$

$$V(\boldsymbol{r}) = \alpha_V \rho_V + \gamma_V \rho_V^3 + \delta_V \Delta \rho_V + e\frac{1-\tau_3}{2}A^0 + \alpha_{TV}\tau_3\rho_3 + \delta_{TV}\tau_3\Delta\rho_3. \tag{4}$$

The pairing potential Δ reads as

$$\Delta(\boldsymbol{r}_1, \boldsymbol{r}_2) = V^{pp}(\boldsymbol{r}_1, \boldsymbol{r}_2)\kappa(\boldsymbol{r}_1, \boldsymbol{r}_2), \tag{5}$$

with the density-dependent zero-range pairing force V^{pp},

$$V^{pp}(\boldsymbol{r}_1, \boldsymbol{r}_2) = \frac{1}{2}V_0(1 - P^\sigma)\delta(\boldsymbol{r}_1 - \boldsymbol{r}_2)\left(1 - \frac{\rho(\boldsymbol{r}_1)}{\rho_{\text{sat}}}\right). \tag{6}$$

and the pairing tensor $\kappa = V^* U^T$.

2.2. The Lipkin–Nogami Method

The Lipkin–Nogami method [28–30] is an approximation to the variation after particle number projection. Its detailed formalism and its comparison with the exact projection method can be found in Ref. [26].

The application of the LN method to the framework of RHB gives a modification of the RHB equation, i.e., the RHB+LN equation [27]

$$\begin{pmatrix} h_D - \lambda - 2\lambda_2(1-2\rho) & \Delta \\ -\Delta^* & -h_D^* + \lambda + 2\lambda_2(1-2\rho^*) \end{pmatrix} \begin{pmatrix} U_k \\ V_k \end{pmatrix} = (E_k - \lambda_2) \begin{pmatrix} U_k \\ V_k \end{pmatrix}, \quad (7)$$

where λ_2 is given by

$$\lambda_2 = -\frac{1}{4} \frac{\mathrm{Tr}_2\mathrm{Tr}_2[(\kappa^*\rho)\bar{v}(\sigma\kappa)]}{[\mathrm{Tr}(\kappa\kappa^\dagger)]^2 - 2\mathrm{Tr}(\kappa\kappa^\dagger\kappa\kappa^\dagger)} \quad (8)$$

with ρ being the density matrix and $\sigma = 1 - \rho$ and $\bar{v}_{abcd} = \langle ab|V^{pp}|cd - dc\rangle$ being the anti-symmetrized matrix elements of the pairing force V^{pp}. The trace Tr_2 means the summation only runs over the particle–particle channel [25].

After the approximate particle number projection, the total energy reads as [49]

$$E_{\mathrm{LN}} = \langle \hat{H} \rangle - \lambda_2 \langle (\Delta \hat{N})^2 \rangle. \quad (9)$$

Similarly, the physical observables should be calculated with the approximately particle number projected state. In particular, the LN occupation probabilities in canonical basis are given by [31,49]

$$w_k = v_k^2 + \overline{u^2v^2} \frac{u_k^2 v_k^2 \overline{u^2v^2(u^2-v^2)} - u_k^2 v_k^2 (u_k^2 - v_k^2)\overline{u^2v^2}}{2(\overline{u^2v^2})^3 + (\overline{u^2v^2})^2 - 6\overline{u^2v^2}\,\overline{u^4v^4} - [\overline{u^2v^2(u^2-v^2)}]^2} \quad (10)$$

with $\bar{x} = \sum_{k>0} x_k$.

3. Numerical Details

For the present investigation of Ca isotopes, the spherical RCHB calculations are adopted. It is noted that in Ref. [50], the DRHBc theory was used to calculate all the even–even nuclei with $8 \leq Z \leq 120$, and the authors found that the investigated Ca isotopes were all spherical. As mentioned above, instead of coordinate space [11], here the RCHB theory is solved in the DWS basis [42]. This is actually a simplification of the DRHBc theory with the assumption of spherical symmetry, i.e., the Hamiltonian matrix is expanded in j^π-blocks instead of m^π-blocks. Based on this, the LN correction is further implemented.

The relativistic density functional PC-PK1 [51] is employed in the particle–hole channel, and in the particle–particle channel, the density-dependent zero-range pairing force (6) is adopted, where the pairing strength $V_0 = -325$ MeV $\cdot$ fm^3 and the saturation density $\rho_{\mathrm{sat}} = 0.152$ fm^{-3} together with a pairing window of 100 MeV. The energy cutoff for the DWS basis in the Fermi sea is $E^+_{\mathrm{cut}} = 300$ MeV. The angular momentum cutoff for the DWS basis is $J_{\mathrm{cut}} = 23/2\,\hbar$. The number of DWS basis states in the Dirac sea is taken to be the same as that in the Fermi sea. All these numerical details are well examined in the DRHBc mass table calculations [50,52–54], where the RHB equation is also solved in a spherical DWS basis.

In the RCHB+LN calculations, all the numerical details are taken to be the same as those in the RCHB calculations except the pairing strength, because the LN correction gives stronger pairing correlations with the same pairing force (cf. Figure 4a,b in Ref. [27]). Figure 1 shows the odd–even mass differences for Ca isotopes calculated by RCHB+LN in comparison with those in RCHB calculations, as well as the corresponding experimental data [55]. It is found that in RCHB+LN calculations, the odd–even mass differences of most Ca isotopes are overestimated when the pairing strength is chosen to be $V_0 = -325$ MeV $\cdot$ fm^3. Therefore, the pairing strength was readjusted to $V_0 = -300$ MeV $\cdot$ fm^3 in order to reproduce the experimental data.

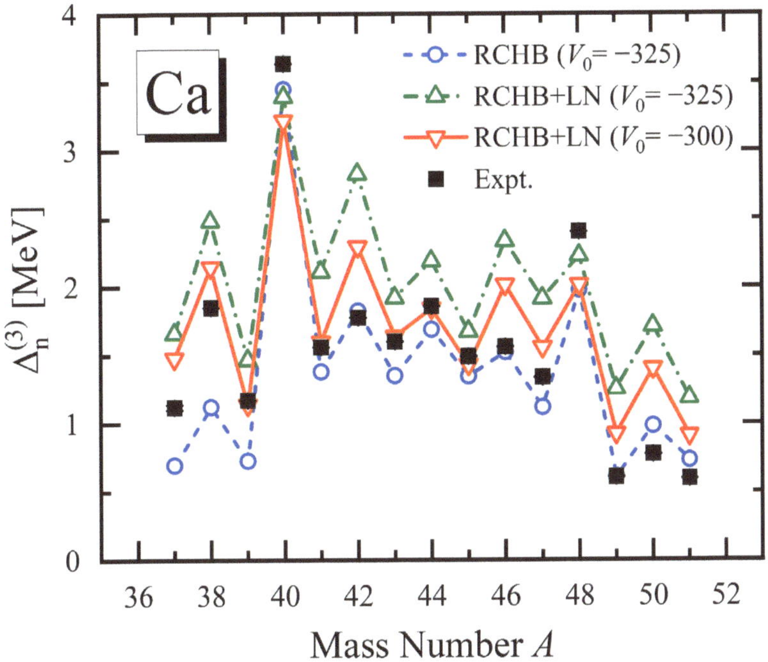

Figure 1. The odd–even differences of Ca isotopic chain in the relativistic continuum Hartree–Bogoliubov+Lipkin–Nogami (RCHB+LN) calculations versus the mass number for $V_0 = -325$ MeV · fm^3 (triangle) and for $V_0 = -300$MeV · fm^3 (inverted triangle). The experimental data [55] (square) and the results in RCHB calculations (circle) are shown for comparison.

4. Results and Discussion

The ground-state properties of even Ca isotopes calculated by RCHB and RCHB+LN are shown in Figure 2. The results of RCHB+LN with $V_0 = -325$ MeV · fm^3 are also shown for comparison. The binding energies per nucleon BE/A are shown in Figure 2a. It is found that larger BE/A are given by including the LN correction, while the trend does not change. The two-neutron separation energies S_{2n} are shown in Figure 2b. In RCHB results, kinks appear at the traditional magic or submagic numbers $N = 20$, 28, and 40. With LN correction, all the amplitudes of these kinks become smaller. At $N = 50$, however, there seems no kink for both the RCHB and RCHB+LN, indicating the disappearance of the traditional magic number $N = 50$ in Ca isotopes. In addition, in the region of $N > 40$, the nuclei become weakly-bound, and the S_{2n} decrease slowly both in the RCHB and RCHB+LN calculations. Such behavior of S_{2n} indicates the appearance of giant halos in the Ca isotopes. Moreover, both the RCHB and the RCHB+LN calculations show the two-neutron drip line of the Ca isotopic chain at $N = 60$. The root mean square (rms) neutron radii in even Ca isotopes are plotted in Figure 2c. The rms radii are nearly unchanged with the LN correction. In addition, from $N = 40$, the rms radii gradually deviate from the empirical $r_0N^{1/3}$ formula as the neutron number increases. This behavior also indicates the possible existence of the halo or giant halo phenomena.

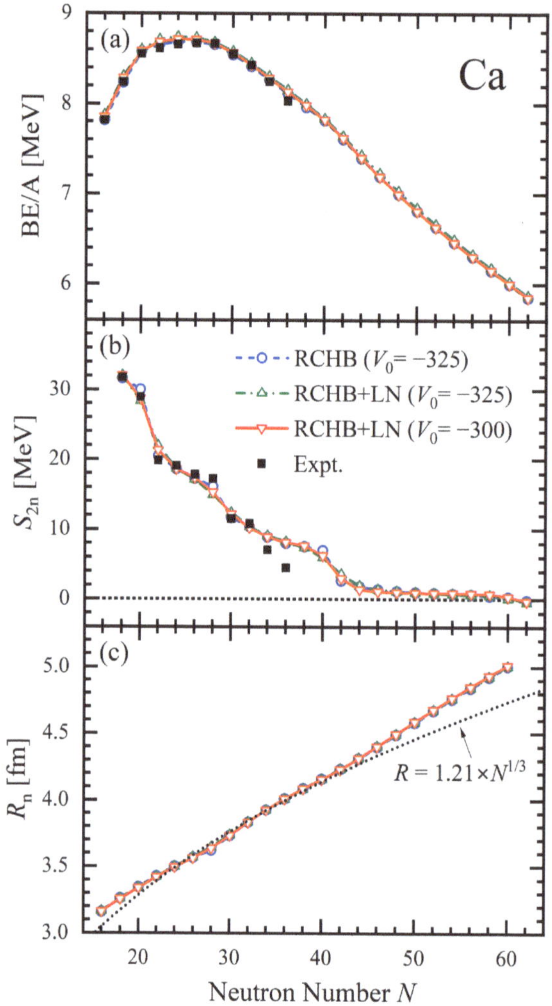

Figure 2. (color online) The ground-state properties of even Ca isotopes, including (**a**) the binding energies per nucleon BE/A, (**b**) the two-neutron separation energies S_{2n}, and (**c**) the root mean square neutron radii R_n obtained from the RCHB and RCHB+LN calculations. The experimental results (square) of BE/A and S_{2n}, as well as the empirical formula (dotted line) of R_n, are also shown for comparison.

Figure 3 shows the neutron density distributions of Ca isotopes derived from RCHB and RCHB+LN calculations. In the region of $N > 40$, the neutron density distribution of Ca isotopes with $A > 62$ are much more diffused than those of ^{60,62}Ca for both the RCHB and RCHB+LN. Combined with the S_{2n} and R_n results mentioned before, this indicates that $^{64-80}$Ca are halo nuclei but that ^{62}Ca is not, and the LN correction does not change the range of those with halo.

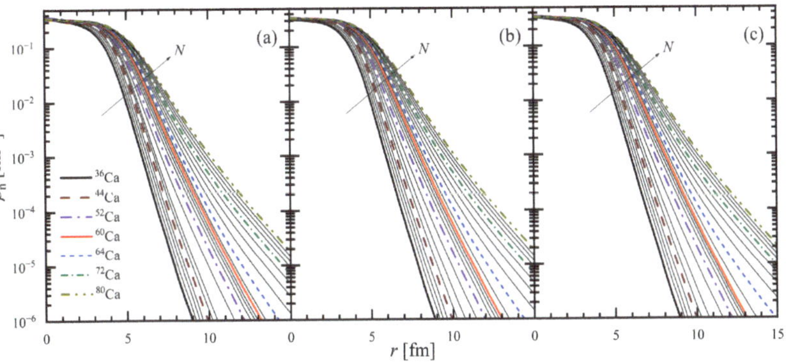

Figure 3. (color online) The neutron density distributions of even Ca isotopes obtained from RCHB+LN for (**b**) $V_0 = -325$ MeV · fm^3 and (**c**) $V_0 = -300$ MeV · fm^3 in comparison with (**a**) RCHB results for $V_0 = -325$ MeV · fm^3.

To see the impact of the LN correction more clearly, as in Refs. [13,14], we take ^{66}Ca as an example. Figure 4 shows its single-neutron levels in canonical basis around the Fermi surface, where the results of the RCHB with $V_0 = -325$ MeV · fm^3, RCHB+LN with $V_0 = -325$ MeV · fm^3, and RCHB+LN with $V_0 = -300$ MeV · fm^3 are plotted in the three panels, respectively. Comparison between these three calculations shows that the LN correction only slightly changes the energies and the occupation probabilities of the levels. For example, the energy and the occupation probability of the neutron $2d_{3/2}$ level with the largest difference are 2.05 MeV and 6.2% for the RCHB with $V_0 = -325$ MeV · fm^3, 1.96 MeV and 7.0% for the RCHB+LN with $V_0 = -325$ MeV · fm^3, and 1.87 MeV and 5.2% for the RCHB+LN with $V_0 = -300$ MeV · fm^3, respectively.

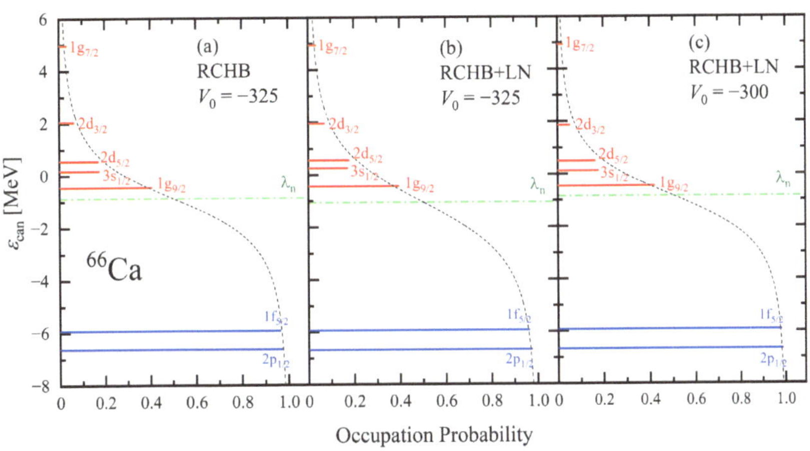

Figure 4. (color online) The single-neutron levels around the Fermi surface in the canonical basis for ^{66}Ca versus the occupation probability (v^2 for RCHB and w for RCHB+LN) obtained from RCHB+LN for (**b**) $V_0 = -325$ MeV · fm^3 and (**c**) $V_0 = -300$ MeV · fm^3 in comparison with (**a**) RCHB results for $V_0 = -325$ MeV · fm^3. The dash-dotted lines indicate the neutron Fermi surfaces. The dotted curves correspond to the BCS or BCS+LN formula with corresponding average pairing gap.

To further check the influence of the LN correction on the potential giant halo in ^{66}Ca, the rms radii of each single-neutron level around the Fermi surface in RCHB and RCHB+LN calculations are compared in Figure 5. The rms radius of the neutron $3s_{1/2}$ level is much larger than other neutron levels nearby both in RCHB and RCHB+LN. This shows that the neutron $3s_{1/2}$ level plays a crucial role in the giant halo in Ca isotopes, as was pointed out in Ref. [13]. With LN correction, when the same pairing strength $V_0 = -325$ MeV $\cdot$ fm^3 is employed, the rms radii of these single-neutron levels do not change much. Among them, the rms radius of the neutron $2d_{5/2}$ level changes most for about 0.04 fm. In contrast, when the pairing strength is readjusted to $V_0 = -300$ MeV $\cdot$ fm^3, although most of the radii do not change much, there are considerable changes in those of the neutron $3s_{1/2}$, $2d_{5/2}$, and $2d_{3/2}$ levels. Especially for neutron $3s_{1/2}$ level, its rms radius increases from 6.775 fm (RCHB, $V_0 = -325$) and 6.804 fm (RCHB+LN, $V_0 = -325$) to 7.100 fm (RCHB+LN, $V_0 = -300$).

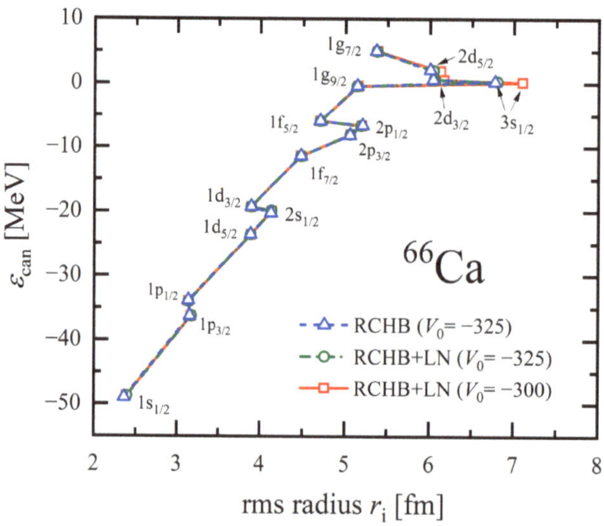

Figure 5. (color online) The single-particle energies in the canonical basis $\varepsilon_{\rm can}$ of each single-neutron level around the Fermi surface in ^{66}Ca versus its root mean square radius r_i.

As shown in Figure 4, there is a large shell gap between the $1g_{9/2}$ level and $1f_{5/2}$ level both in the RCHB and RCHB+LN, which corresponds to the submagic number $N = 40$. Thus, one can naturally take the levels below the $N = 40$ shell gap as the "core", and the remaining weakly bound levels and the levels in continuum above the shell gap as the "halo". Correspondingly, one can decompose the neutron density $\rho_n(\boldsymbol{r})$ into the "core" part and the "halo" part, as is plotted in Figure 6. The three panels show the results of the RCHB with $V_0 = -325$ MeV $\cdot$ fm^3, RCHB+LN with $V_0 = -325$ MeV $\cdot$ fm^3, and RCHB+LN with $V_0 = -300$ MeV $\cdot$ fm^3, respectively. The halo part surpasses the core part at about $r = 8$ fm and gradually domains with increasing r in all three calculations, indicating the halo structure in ^{66}Ca. The neutron numbers in the halo are 6.45 for both the RCHB and RCHB+LN with $V_0 = -325$ MeV $\cdot$ fm^3 and 6.25 for the RCHB+LN with $V_0 = -300$ MeV $\cdot$ fm^3. The results show that the LN correction does not change the appearance of giant halos in ^{66}Ca. On the one hand, under the same pairing strength, both the neutron number in the halo and the neutron density distribution of RCHB+LN are nearly unchanged in comparison with that of RCHB. On the other hand, with the pairing strength readjusted, the neutron density distribution of the RCHB+LN becomes slightly more diffused, although the neutron number in the halo becomes a little smaller.

To further analyze the neutron density, the density distributions of several single-neutron levels in the halo are also depicted in Figure 6. The neutron $3s_{1/2}$ level contributes to the halo most both in the RCHB and RCHB+LN. This can be seen more clearly in Figure 7, where the relative contributions to the total neutron density of these single-neutron levels are plotted. At $r = 15$ fm, the relative contribution of the neutron $3s_{1/2}$ level is about 50% in the RCHB with $V_0 = -325$ MeV $\cdot$ fm^3, about 48% in the RCHB+LN with $V_0 = -325$ MeV $\cdot$ fm^3, and about 55% in the RCHB+LN with $V_0 = -300$ MeV $\cdot$ fm^3. Such a result shows that with LN correction, when the pairing strength is $V_0 = -325$ MeV $\cdot$ fm^3, the neutron $3s_{1/2}$ level contributes slightly less to the giant halo in ^{66}Ca. In contrast, when the pairing strength is readjusted to $V_0 = -300$ MeV $\cdot$ fm^3, it contributes more. As the density distribution of neutron $3s_{1/2}$ level is more diffused than other levels nearby (cf. Figure 6), it is no wonder why the total neutron density becomes more diffused with LN correction and the readjusted pairing strength.

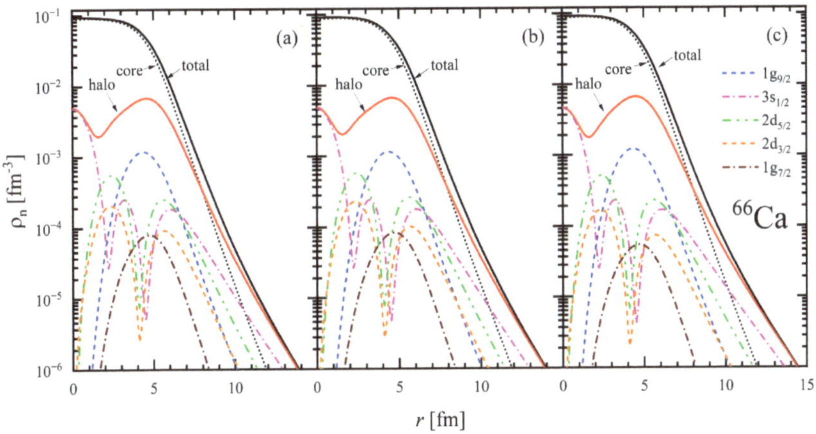

Figure 6. (color online) The total neutron density distribution of ^{66}Ca, as well as its decomposition into the core and the halo, obtained from RCHB+LN for (**b**) $V_0 = -325$ MeV $\cdot$ fm^3 and (**c**) $V_0 = -300$ MeV $\cdot$ fm^3 in comparison with (**a**) RCHB results for $V_0 = -325$ MeV $\cdot$ fm^3. Contributions from several single-neutron levels in the halo are also given.

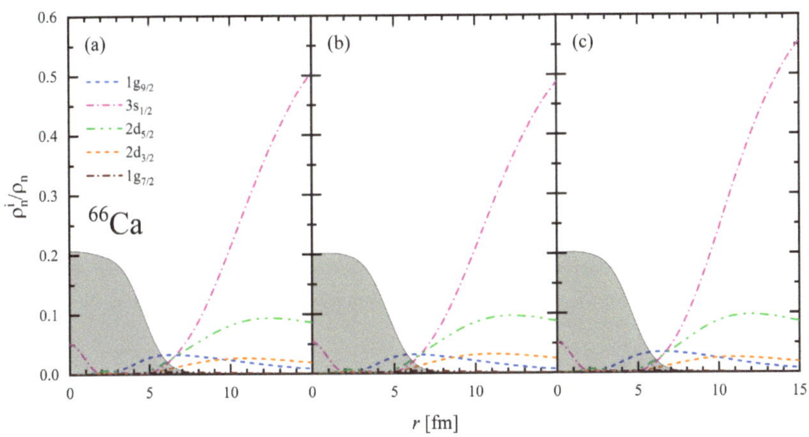

Figure 7. (color online) The relative contributions of several single-neutron levels in the halo of ^{66}Ca obtained from RCHB+LN for (**b**) $V_0 = -325$ MeV $\cdot$ fm^3 and (**c**) $V_0 = -300$ MeV $\cdot$ fm^3 in comparison with (**a**) RCHB results for $V_0 = -325$ MeV $\cdot$ fm^3. The shade areas indicate the total neutron density in arbitrary units.

5. Conclusions

In this paper, the impact of the Lipkin–Nogami (LN) method on giant halos has been investigated within the relativistic continuum Hartree–Bogoliubov (RCHB) theory. Here, the RHB and RHB+LN equations were solved in a spherical Dirac Woods–Saxon basis. Reproducing the odd–even mass differences of Ca isotopes, the pairing strength was readjusted to $V_0 = -300$ MeV $\cdot$ fm^3 for RCHB+LN. Comparing the ground-state properties of Ca isotopes calculated by the RCHB with $V_0 = -325$ MeV $\cdot$ fm^3, the RCHB+LN with $V_0 = -325$ MeV $\cdot$ fm^3, and the RCHB+LN with $V_0 = -300$ MeV $\cdot$ fm^3, this shows that the LN correction does not change the range of Ca isotopes with giant halos. Taking ^{66}Ca as an example, the neutron density distribution obtained with LN correction and readjusted pairing strength has been found to be slightly more diffused than that of the RCHB, which can be illustrated by the enlargement of the root mean square radius and the enhancement of the relative contribution for neutron $3s_{1/2}$ level.

Note that present results are limited to spherical halo nuclei, and the influence of the LN correction on the deformed halo nuclei is a topic to be investigated in the future. As the LN method is an approximation to the exact particle number projection method, how the exact particle number projection works on halo nuclei is an interesting topic. Besides halo phenomenon, neutron skin has also attracted much attention [56–58], and further investigation on it can be carried out. In addition, studies on odd nuclei are also important. For instance, in a recent work [48], the even–odd staggering of charge radii in Hg isotopes has been studied by the DRHBc theory and attributed to shape staggering. The even–odd staggering of charge radii in Ca isotopes is also interesting, which has attracted many theoretical efforts [59–63], and thus future work about it based on the DRHBc theory is welcome.

Author Contributions: Conceptualization, C.Z.; methodology, C.Z.; software, C.Z.; formal analysis, All authors; investigation, C.Z. and P.G.; writing—original draft preparation, C.Z.; writing—review and editing, All authors; visualization, C.Z.; project administration, X.J. All authors have read and agreed to the published version of the manuscript.

Funding: This work was partly supported by the National Natural Science Foundation of China under Grants No. 11935003, No. 12435006, No. 12141501, and No. 12475117; the State Key Laboratory of Nuclear Physics and Technology; Peking University under Grant No. NPT2023ZX03; and the National Key Laboratory of Neutron Science and Technology under Grant No. NST202401016.

Data Availability Statement: The dataset can be accessed upon request to the corresponding author.

Acknowledgments: The authors are indebted to J. Meng and S.-Q. Zhang for their constructive guidance and valuable suggestions. This work was supported by High-performance Computing Platform of Peking University. Helpful discussions with K.-Y. Zhang, C. Pan, and the members of the DRHBc Mass Table Collaboration are highly appreciated.

Conflicts of Interest: The authors declare no conflicts of interest.

References

1. Vretenar, D.; Afanasjev, A.; Lalazissis, G.; Ring, P. Relativistic Hartree-Bogoliubov theory: Static and dynamic aspects of exotic nuclear structure. *Phys. Rep.* **2005**, *409*, 101–259. [CrossRef]
2. Gade, A.; Glasmacher, T. In-beam nuclear spectroscopy of bound states with fast exotic ion beams. *Prog. Part. Nucl. Phys.* **2008**, *60*, 161–224. [CrossRef]
3. Otsuka, T.; Gade, A.; Sorlin, O.; Suzuki, T.; Utsuno, Y. Evolution of shell structure in exotic nuclei. *Rev. Mod. Phys.* **2020**, *92*, 015002. [CrossRef]
4. Yamaguchi, T.; Koura, H.; Litvinov, Y.; Wang, M. Masses of exotic nuclei. *Prog. Part. Nucl. Phys.* **2021**, *120*, 103882. [CrossRef]
5. Tanihata, I.; Hamagaki, H.; Hashimoto, O.; Shida, Y.; Yoshikawa, N.; Sugimoto, K.; Yamakawa, O.; Kobayashi, T.; Takahashi, N. Measurements of Interaction Cross Sections and Nuclear Radii in the Light p-Shell Region. *Phys. Rev. Lett.* **1985**, *55*, 2676–2679. [CrossRef]
6. Zhang, K.Y.; Papakonstantinou, P.; Mun, M.H.; Kim, Y.; Yan, H.; Sun, X.X. Collapse of the $N = 28$ shell closure in the newly discovered ^{39}Na nucleus and the development of deformed halos towards the neutron dripline. *Phys. Rev. C* **2023**, *107*, 1041303. [CrossRef]

7. Kobayashi, T.; Yamakawa, O.; Omata, K.; Sugimoto, K.; Shimoda, T.; Takahashi, N.; Tanihata, I. Projectile Fragmentation of the Extremely Neutron-Rich Nucleus ^{11}Li at 0.79 GeV/nucleon. *Phys. Rev. Lett.* **1988**, *60*, 2599–2602. [CrossRef] [PubMed]
8. Orr, N.A.; Anantaraman, N.; Austin, S.M.; Bertulani, C.A.; Hanold, K.; Kelley, J.H.; Morrissey, D.J.; Sherrill, B.M.; Souliotis, G.A.; Thoennessen, M.; et al. Momentum distributions of ^{9}Li fragments following the breakup of ^{11}Li. *Phys. Rev. Lett.* **1992**, *69*, 2050–2053. [CrossRef]
9. Ieki, K.; Sackett, D.; Galonsky, A.; Bertulani, C.A.; Kruse, J.J.; Lynch, W.G.; Morrissey, D.J.; Orr, N.A.; Schulz, H.; Sherrill, B.M.; et al. Coulomb dissociation of ^{11}Li. *Phys. Rev. Lett.* **1993**, *70*, 730–733. [CrossRef]
10. Meng, J.; Ring, P. Relativistic Hartree-Bogoliubov Description of the Neutron Halo in ^{11}Li. *Phys. Rev. Lett.* **1996**, *77*, 3963–3966. [CrossRef] [PubMed]
11. Meng, J. Relativistic continuum Hartree-Bogoliubov theory with both zero range and finite range Gogny force and their application. *Nucl. Phys. A* **1998**, *635*, 3–42. [CrossRef]
12. Meng, J.; Ring, P. Giant Halo at the Neutron Drip Line. *Phys. Rev. Lett.* **1998**, *80*, 460–463. [CrossRef]
13. Meng, J.; Toki, H.; Zeng, J.Y.; Zhang, S.Q.; Zhou, S.G. Giant halo at the neutron drip line in Ca isotopes in relativistic continuum Hartree-Bogoliubov theory. *Phys. Rev. C* **2002**, *65*, 041302. [CrossRef]
14. Terasaki, J.; Zhang, S.Q.; Zhou, S.G.; Meng, J. Giant halos in relativistic and nonrelativistic approaches. *Phys. Rev. C* **2006**, *74*, 054318. [CrossRef]
15. Sandulescu, N.; Geng, L.S.; Toki, H.; Hillhouse, G.C. Pairing correlations and resonant states in the relativistic mean field theory. *Phys. Rev. C* **2003**, *68*, 054323. [CrossRef]
16. Zhang, S.S.; Xu, X.D.; Peng, J.P. Hints of giant halo in Zr isotopes by resonant RMF+ACCC+BCS approach. *Eur. Phys. J. A* **2012**, *48*. [CrossRef]
17. Ding, K.M.; Shi, M.; Guo, J.Y.; Niu, Z.M.; Liang, H. Resonant-continuum relativistic mean-field plus BCS in complex momentum representation. *Phys. Rev. C* **2018**, *98*, 014316. [CrossRef]
18. Cao, X.N.; Ding, K.M.; Shi, M.; Liu, Q.; Guo, J.Y. Exploration of the exotic structure in Ce isotopes by the relativistic point-coupling model combined with complex momentum representation. *Phys. Rev. C* **2020**, *102*, 044313. [CrossRef]
19. Long, W.H.; Ring, P.; Meng, J.; Van Giai, N.; Bertulani, C.A. Nuclear halo structure and pseudospin symmetry. *Phys. Rev. C* **2010**, *81*, 031302. [CrossRef]
20. Grasso, M.; Yoshida, S.; Sandulescu, N.; Van Giai, N. Giant neutron halos in the non-relativistic mean field approach. *Phys. Rev. C* **2006**, *74*, 064317. [CrossRef]
21. Zhang, Y.; Matsuo, M.; Meng, J. Pair correlation of giant halo nuclei in continuum Skyrme-Hartree-Fock-Bogoliubov theory. *Phys. Rev. C* **2012**, *86*, 054318. [CrossRef]
22. Qu, X.Y.; Zhang, Y. Canonical states in continuum Skyrme Hartree-Fock-Bogoliubov theory with Green's function method. *Phys. Rev. C* **2019**, *99*, 014314. [CrossRef]
23. Sun, T.T. Green's function method in covariant density functional theory. *Sci. Sin.-Phys. Mech. Astron.* **2015**, *46*, 012006. (In Chinese) [CrossRef]
24. Qu, X.Y.; Tong, H.; Zhang, S.Q. Canonical states in relativistic continuum theory with the Green's function method: Neutrons in continuum of zirconium giant-halo nuclei. *Phys. Rev. C* **2022**, *105*, 014326. [CrossRef]
25. Ring, P.; Schuck, P. *The Nuclear Many-Body Problem*; Springer: Berlin, Germany, 2004.
26. Sheikh, J.A.; Dobaczewski, J.; Ring, P.; Robledo, L.M.; Yannouleas, C. Symmetry restoration in mean-field approaches. *J. Phys. G Nucl. Part. Phys.* **2021**, *48*, 123001. [CrossRef]
27. Afanasjev, A.; Ring, P.; König, J. Cranked relativistic Hartree–Bogoliubov theory: Formalism and application to the superdeformed bands in the A∼190 region. *Nucl. Phys. A* **2000**, *676*, 196–244. [CrossRef]
28. Lipkin, H.J. Collective motion in many-particle systems. *Ann. Phys.* **1960**, *9*, 272–291. [CrossRef]
29. Nogami, Y. Improved Superconductivity Approximation for the Pairing Interaction in Nuclei. *Phys. Rev.* **1964**, *134*, B313–B321. [CrossRef]
30. Pradhan, H.; Nogami, Y.; Law, J. Study of approximations in the nuclear pairing-force problem. *Nucl. Phys. A* **1973**, *201*, 357–368. [CrossRef]
31. Bennour, L.; Heenen, P.H.; Bonche, P.; Dobaczewski, J.; Flocard, H. Charge distributions of ^{208}Pb, ^{206}Pb, and ^{205}Tl and the mean-field approximation. *Phys. Rev. C* **1989**, *40*, 2834–2839. [CrossRef]
32. Shi, H.; Wang, X.B.; Dong, G.X.; Wang, H. Abnormal odd-even staggering behavior around ^{132}Sn studied by density functional theory. *Chin. Phys. C* **2020**, *44*, 094108. [CrossRef]
33. Raj, R.; Rustgi, M.L. Modified Tamm-Dancoff approximation calculations using Lipkin-Nogami approach. *Phys. Rev. C* **1982**, *26*, 243–248. [CrossRef]
34. Raj, R.; Rustgi, M.L. Odd Ni isotopes in the Lipkin-Nogami approach. *Phys. Rev. C* **1983**, *28*, 935–937. [CrossRef]
35. Terasaki, J.; Heenen, P.H.; Bonche, P.; Dobaczewski, J.; Flocard, H. Superdeformed rotational bands with density dependent pairing interactions. *Nucl. Phys. A* **1995**, *593*, 1–20. [CrossRef]
36. Afanasjev, A.V.; König, J.; Ring, P. Cranked relativistic Hartree-Bogoliubov theory: Superdeformed bands in the $A \sim 190$ region. *Phys. Rev. C* **1999**, *60*, 051303. [CrossRef]
37. Hung, N.Q.; Dang, N.D. Canonical and microcanonical ensemble descriptions of thermal pairing within BCS and quasiparticle random-phase approximation. *Phys. Rev. C* **2010**, *81*, 057302. [CrossRef]

38. Saleh Yousef, M.; Elsharkawy, H.; Rashed, N. Matrix elements of the two-neutrino double beta decay of ^{76}Ge using deformed BCS and Lipkin–Nogami approaches. *Nucl. Phys. A* **2018**, *975*, 97–106. [CrossRef]
39. Meng, J.; Toki, H.; Zhou, S.; Zhang, S.; Long, W.; Geng, L. Relativistic continuum Hartree Bogoliubov theory for ground-state properties of exotic nuclei. *Prog. Part. Nucl. Phys.* **2006**, *57*, 470–563. [CrossRef]
40. Meng, J.; Zhou, S.G. Halos in medium-heavy and heavy nuclei with covariant density functional theory in continuum. *J. Phys. G: Nucl. Part. Phys.* **2015**, *42*, 093101. [CrossRef]
41. Kucharek, H.; Ring, P. Relativistic field theory of superfluidity in nuclei. *Z. Phys. A* **1991**, *339*, 23–35. [CrossRef]
42. Zhou, S.G.; Meng, J.; Ring, P. Spherical relativistic Hartree theory in a Woods-Saxon basis. *Phys. Rev. C* **2003**, *68*, 034323. [CrossRef]
43. Zhou, S.G.; Meng, J.; Ring, P.; Zhao, E.G. Neutron halo in deformed nuclei. *Phys. Rev. C* **2010**, *82*, 011301. [CrossRef]
44. Sun, X.X.; Zhao, J.; Zhou, S.G. Shrunk halo and quenched shell gap at $N = 16$ in ^{22}C: Inversion of *sd* states and deformation effects. *Phys. Lett. B* **2018**, *785*, 530–535. [CrossRef]
45. Zhang, K.; He, X.; Meng, J.; Pan, C.; Shen, C.; Wang, C.; Zhang, S. Predictive power for superheavy nuclear mass and possible stability beyond the neutron drip line in deformed relativistic Hartree-Bogoliubov theory in continuum. *Phys. Rev. C* **2021**, *104*, l021301. [CrossRef]
46. Choi, Y.B.; Lee, C.H.; Mun, M.H.; Kim, Y. Bubble nuclei with shape coexistence in even-even isotopes of Hf to Hg. *Phys. Rev. C* **2022**, *105*, 024306. [CrossRef]
47. Guo, P.; Pan, C.; Zhao, Y.C.; Du, X.K.; Zhang, S.Q. Prolate-shape dominance in atomic nuclei within the deformed relativistic Hartree-Bogoliubov theory in continuum. *Phys. Rev. C* **2023**, *108*, 014319. [CrossRef]
48. Mun, M.H.; Kim, S.; Cheoun, M.K.; So, W.; Choi, S.; Ha, E. Odd-even shape staggering and kink structure of charge radii of Hg isotopes by the deformed relativistic Hartree–Bogoliubov theory in continuum. *Phys. Lett. B* **2023**, *847*, 138298. [CrossRef]
49. Nikšić, T.; Vretenar, D.; Ring, P. Beyond the relativistic mean-field approximation. II. Configuration mixing of mean-field wave functions projected on angular momentum and particle number. *Phys. Rev. C* **2006**, *74*, 064309. [CrossRef]
50. Zhang, K.; Cheoun, M.K.; Choi, Y.B.; Chong, P.S.; Dong, J.; Dong, Z.; Du, X.; Geng, L.; Ha, E.; He, X.T.; et al. Nuclear mass table in deformed relativistic Hartree–Bogoliubov theory in continuum, I: Even–even nuclei. *At. Data Nucl. Data Tables* **2022**, *144*, 101488. [CrossRef]
51. Zhao, P.W.; Li, Z.P.; Yao, J.M.; Meng, J. New parametrization for the nuclear covariant energy density functional with a point-coupling interaction. *Phys. Rev. C* **2010**, *82*, 054319. [CrossRef]
52. Zhang, K.; Cheoun, M.K.; Choi, Y.B.; Chong, P.S.; Dong, J.; Geng, L.; Ha, E.; He, X.; Heo, C.; Ho, M.C.; et al. Deformed relativistic Hartree-Bogoliubov theory in continuum with a point-coupling functional: Examples of even-even Nd isotopes. *Phys. Rev. C* **2020**, *102*, 024314. [CrossRef]
53. Pan, C.; Cheoun, M.K.; Choi, Y.B.; Dong, J.; Du, X.; Fan, X.H.; Gao, W.; Geng, L.; Ha, E.; He, X.T.; et al. Deformed relativistic Hartree-Bogoliubov theory in continuum with a point-coupling functional. II. Examples of odd Nd isotopes. *Phys. Rev. C* **2022**, *106*, 014316. [CrossRef]
54. Guo, P.; Cao, X.; Chen, K.; Chen, Z.; Cheoun, M.K.; Choi, Y.B.; Lam, P.C.; Deng, W.; Dong, J.; Du, P.; et al. Nuclear mass table in deformed relativistic Hartree–Bogoliubov theory in continuum, II: Even-Z nuclei. *At. Data Nucl. Data Tables* **2024**, *158*, 101661. [CrossRef]
55. Wang, M.; Huang, W.; Kondev, F.; Audi, G.; Naimi, S. The AME 2020 atomic mass evaluation (II). Tables, graphs and references. *Chin. Phys. C* **2021**, *45*, 030003. [CrossRef]
56. Centelles, M.; Roca-Maza, X.; Viñas, X.; Warda, M. Origin of the neutron skin thickness of ^{208}Pb in nuclear mean-field models. *Phys. Rev. C* **2010**, *82*, 054314. [CrossRef]
57. Roca-Maza, X.; Centelles, M.; Viñas, X.; Warda, M. Neutron Skin of ^{208}Pb, Nuclear Symmetry Energy, and the Parity Radius Experiment. *Phys. Rev. Lett.* **2011**, *106*, 252501. [CrossRef] [PubMed]
58. Warda, M.; Centelles, M.; Viñas, X.; Roca-Maza, X. Influence of the single-particle structure on the nuclear surface and the neutron skin. *Phys. Rev. C* **2014**, *89*, 064302. [CrossRef]
59. Fayans, S.; Tolokonnikov, S.; Trykov, E.; Zawischa, D. Nuclear isotope shifts within the local energy-density functional approach. *Nucl. Phys. A* **2000**, *676*, 49–119. [CrossRef]
60. Reinhard, P.G.; Nazarewicz, W. Toward a global description of nuclear charge radii: Exploring the Fayans energy density functional. *Phys. Rev. C* **2017**, *95*, 064328. [CrossRef]
61. Miller, A.J.; Minamisono, K.; Klose, A.; Garand, D.; Kujawa, C.; Lantis, J.D.; Liu, Y.; Maaß, B.; Mantica, P.F.; Nazarewicz, W.; et al. Proton superfluidity and charge radii in proton-rich calcium isotopes. *Nat. Phys.* **2019**, *15*, 432–436. [CrossRef]
62. An, R.; Geng, L.S.; Zhang, S.S. Novel ansatz for charge radii in density functional theories. *Phys. Rev. C* **2020**, *102*, 024307. [CrossRef]
63. Yang, D.; Rong, Y.T.; An, R.; Shi, R.X. Potential signature of new magicity from universal aspects of nuclear charge radii. *arXiv* **2024**, arXiv:2411.03076. [CrossRef]

particles

MDPI

Article

Bubble Structure in Isotopes of Lu to Hg

Changhoon Song [1,†], Yongbeom Choi [2,†], Youngman Kim [3] and Chang-Hwan Lee [1,*]

[1] Department of Physics, Pusan National University, Busan 46241, Republic of Korea; wtsch0221@gmail.com
[2] School of Physics, Peng Huanwu Collaborative Center for Research and Education, and International Research Center for Big-Bang Cosmology and Element Genesis, Beihang University, Beijing 100191, China; 1991.yb.choi@gmail.com
[3] Center for Exotic Nuclear Studies, Institute for Basic Science, Daejeon 34126, Republic of Korea; ykim@ibs.re.kr
[*] Correspondence: clee@pusan.ac.kr; Tel.: +82-10-9720-8124
[†] These authors contributed equally to this work.

Abstract: Bubble nuclei, characterized by a depletion in nucleon density at the nuclear center, are investigated within the atomic number range $71 \leq Z \leq 80$ using the Deformed Relativistic Hartree–Bogoliubov theory in continuum. This study extends previous investigations, which were limited to even–even isotopes, by incorporating even–odd, odd–even, and odd–odd nuclei within this range. The extension is achieved by introducing the blocking effect into the point-coupling approach to ensure self-consistency. Following previous studies, we define a nucleus as a bubble candidate if the bubble parameter exceeds $\mathcal{B}_p^\star = 20\%$, and identify five bubble nuclei in both even-Z and odd-Z nuclei groups, based on the highest $\mathcal{B}_p^\star$ values. The formation of bubble structures is confirmed through an analysis of proton single-particle energy levels of the most centrally depleted nuclei across four categories: even–odd, even–even, odd–even, and odd–odd.

Keywords: bubble nuclei; neutron drip line

1. Introduction

The exploration of atomic nuclei, particularly those near the neutron drip line, has progressed significantly in recent years, driven by advancements in experimental facilities worldwide [1,2]. The discovery of over 3000 isotopes has broadened our understanding of the nuclear landscape, revealing previously uncharted regions of a nuclear structure. These findings play a crucial role in deepening our knowledge of the nuclear structure and the evolution of heavy nuclei. However, despite these advancements, the evolution of nuclear density distributions remains incompletely understood, particularly regarding the shape decoupling effect, which leads to distinct deformations between the core and the halo, and the existence of bound nuclei beyond the neutron drip line.

A key issue in nuclear structure research is understanding exotic nuclei, which exhibit unique phenomena such as bubble structures and shape coexistence in their ground states [3]. Bubble nuclei, in particular, arise when the central region of the nucleus exhibits a substantial depletion in nucleon density. These nuclei, characterized by reduced proton and neutron densities, have been extensively studied theoretically. It has been suggested that bubble formation originates from the reduced occupation of s states near the Fermi surface, which suppresses the central density [3–12]. Additionally, pairing correlations and nuclear deformation are known to further weaken the bubble structure [3,6,13–17].

In this study, we employ the deformed relativistic Hartree–Bogoliubov theory in continuum (DRHBc) to investigate the nuclear density distributions and internal structure.

This framework is well suited for studying exotic nuclear phenomena, such as neutron-rich nuclei and bubble structures, as it incorporates deformation and treats self-consistently continuum effects and pairing [18–20].

This paper is organized as follows. In Section 2, we introduce the DRHBc framework and discuss its application to bubble structures. Section 3 presents our results on bubble nuclei with atomic numbers ranging $71 \leq Z \leq 80$, identifying those with the most significant central density depletion and analyzing their structural properties. Finally, a conclusion of our findings is provided in Section 4.

2. Theoretical Framework

2.1. DRHBc

In this study, we employ the relativistic mean-field (RMF) framework to investigate nuclear structure and density distributions while incorporating pairing correlations. The blocking effect is included to account for the presence of unpaired nucleons in odd-mass and odd–odd nuclei. This modifies the pairing potential and alters single-particle occupation probabilities, which in turn affect central density depletion in bubble structures. The Lagrangian density used in our calculations is given by [21]

$$
\begin{aligned}
\mathcal{L} = &\bar{\psi}(i\gamma^\mu\partial_\mu - m)\psi - \frac{1}{2}\alpha_S(\bar{\psi}\psi)(\bar{\psi}\psi) - \frac{1}{2}\alpha_V(\bar{\psi}\gamma_\mu\psi)(\bar{\psi}\gamma^\mu\psi) \\
&- \frac{1}{2}\alpha_{TV}(\bar{\psi}\vec{\tau}\gamma_\mu\psi)(\bar{\psi}\vec{\tau}\gamma^\mu\psi) - \frac{1}{3}\beta_S(\bar{\psi}\psi)^3 - \frac{1}{4}\gamma_S(\bar{\psi}\psi)^4 \\
&- \frac{1}{4}\gamma_V\left[(\bar{\psi}\gamma_\mu\psi)(\bar{\psi}\gamma^\mu\psi)\right]^2 - \frac{1}{2}\delta_S\partial_\nu(\bar{\psi}\psi)\partial^\nu(\bar{\psi}\psi) \\
&- \frac{1}{2}\delta_V\partial_\nu(\bar{\psi}\gamma_\mu\psi)\partial^\nu(\bar{\psi}\gamma^\mu\psi) - \frac{1}{2}\delta_{TV}\partial_\nu(\bar{\psi}\vec{\tau}\gamma_\mu\psi)\partial^\nu(\bar{\psi}\vec{\tau}\gamma^\mu\psi) \\
&- \frac{1}{4}F^{\mu\nu}F_{\mu\nu} - e\frac{(1-\tau_3)}{2}\bar{\psi}\gamma^\mu\psi A_\mu,
\end{aligned}
\tag{1}
$$

where m is the nucleon mass, and α_S, α_V, and α_{TV} are the coupling constants for four-fermion contact interactions. The terms with β_S, γ_S, and γ_V account for density-dependent effects, while those with δ_S, δ_V, and δ_{TV} describe the finite-range effects. A_μ and $F^{\mu\nu}$ represent the four-vector potential and the field strength tensor of the electromagnetic field, respectively. The subscripts S, V, and TV denote scalar, vector, and isovector components. Further details on the DRHBc theory can be found in Refs. [19,20], and numerical implementations are described in Ref. [21].

Both automatic and orbital-fixed blocking methods are employed in the DRHBc framework. The automatic method iteratively excludes the lowest-energy quasiparticle state, ensuring self-consistency during convergence. The orbital-fixed method, on the other hand, enforces the occupation of a specific single-particle state. These approaches enable reliable calculations for odd-mass and odd–odd nuclei. Applying the mean-field approximation to the Lagrangian density and performing a Legendre transformation, we derive the mean-field Hamiltonian density. By employing the variational method and the Bogoliubov transformation, we obtain the relativistic Hartree–Bogoliubov (RHB) equation [22], given by

$$
\begin{pmatrix} h_D - \lambda & \Delta \\ -\Delta^* & -h_D^* + \lambda \end{pmatrix} \begin{pmatrix} U_k \\ V_k \end{pmatrix} = E_k \begin{pmatrix} U_k \\ V_k \end{pmatrix},
\tag{2}
$$

where h_D is the Dirac Hamiltonian, λ is the chemical potential, Δ represents the pairing potential, and U_k, V_k are the quasiparticle wavefunctions.

To account for nuclear deformation, the DRHBc framework expands the density distribution in terms of Legendre polynomials as

$$f(\mathbf{r}) = \sum_{\lambda} f_{\lambda}(r) P_{\lambda}(\cos\theta), \quad \lambda = 0, 2, 4, \ldots. \tag{3}$$

where $f_{\lambda}(r)$ are the expansion coefficients, and $P_{\lambda}(\cos\theta)$ are the Legendre polynomials.

In odd-mass and odd–odd nuclei, the blocking effect changes the pairing field by excluding specific quasiparticle states. This leads to modifications in single-particle occupation probabilities and the energy spectrum. The automatic blocking method is particularly effective in cases where a well-defined lowest quasiparticle state exists, facilitating rapid convergence. However, when several low-energy quasiparticle states are closely spaced, the orbital-fixed blocking method is used to investigate the impact of blocking different states on nuclear properties. This ensures a robust and reliable description of pairing suppression and structural evolution in exotic nuclei [20].

2.2. Bubble Parameter

Proton bubble nuclei are identified by a significant reduction in central proton density relative to the maximum proton density within the nucleus. The central density depletion is quantified using the proton depletion fraction, which serves as a key parameter for characterizing bubble structures. In previous studies [6,8,9,14], the bubble parameter was defined as the ratio of the maximum proton density, which is uniquely defined and does not depend on the direction. However, for deformed nuclei described within the DRHBc framework, the proton density distribution varies with the polar angle θ, leading to anisotropic density profiles. Consequently, the maximum proton density also depends on θ, making the original definition unsuitable for deformed nuclei.

To address this issue, a modified bubble parameter $\mathcal{B}_p^{\star}$ was introduced in the previous study [3]. This incorporates an angle-averaged maximum proton density, providing a more consistent measure of central density depletion in deformed nuclei:

$$\mathcal{B}_p^{\star} \equiv \left(1 - \frac{\rho_{p,c}}{\bar{\rho}_{p,\max}}\right) \times 100\ [\%], \tag{4}$$

where the angle averaged maximum proton density $\bar{\rho}_{p,\max}$ is defined as

$$\bar{\rho}_{p,\max} = \frac{\int \rho_p(r,\theta)\delta(r - r_{\max}(\theta))dV}{\int \delta(r - r_{\max}(\theta))dV}. \tag{5}$$

Here, $r_{\max}(\theta)$ denotes the radial coordinate where the proton density reaches its local maximum for a given θ. This formulation ensures that $\bar{\rho}_{p,\max}$ represents an effective maximum density, accounting for directional variations in deformed nuclei. For spherical nuclei, where the proton density distribution is isotropic, the bubble parameter $\mathcal{B}_p^{\star}$ naturally reduces to its conventional definition, i.e., $\mathcal{B}_p^{\star} = \mathcal{B}_p$, ensuring consistency with the previous studies.

3. Results

Using Equation (4), we systematically calculate the bubble parameter for nuclei with $71 \leq Z \leq 80$ described in DRHBc and analyze the result. Table 1 represents the list of nuclei, which show the highest 5 bubble parameter $\mathcal{B}_p^{\star}$ for even-Z and odd-Z nuclei, respectively. We adopt the point coupling density functional (PC-PK1) [23] to extend and compare the previous study. The neutron numbers of the nuclei in Table 1 are close to the neutron magic number 184, with the bubble parameter exceeding 28%. Compared to previous study [3], which identified Hf as the most depleted nucleus, these results reveal that the presence of unpaired nucleons, particularly in odd-Z or odd-N nuclei, significantly contributes to the bubble parameter. In other words, this finding aligns with earlier studies, which indicates that paired nucleons tend to limit the bubble structure. For even-Z nuclei, the nucleus with the most central depletion is Hg, while for odd-Z nuclei, Lu exhibits the most central depletion.

Additionally, we observe the effect of quadrupole deformation (β_2) on the bubble parameter. For even-Z nuclei, a comparison between ^{258}W and ^{253}W indicates that ^{253}W exhibits a smaller bubble parameter. Note that ^{258}W has a magic neutron number, but ^{253}W has a non-magic neutron number. Similarly, for odd-Z nuclei, a comparison between ^{257}Ta and ^{256}Ta shows that ^{256}Ta, with an odd neutron number, has a lower value. These effects can be attributed to the quadrupole deformation (β_2), which reduces the bubble parameter. Additionally, the presence of an unpaired nucleon can alter pairing correlations and subsequently influence the bubble parameter, consistent with previous studies [3,6,13–17].

Table 1. List of the nuclei which show the highest 5 bubble parameters B_p^* for even-Z and odd-Z nuclei, respectively.

Even-Z Nuclei			Odd-Z Nuclei		
Nuclei	β_2	B_p^*[%]	Nuclei	β_2	B_p^*[%]
^{263}Hg	0.000	34.6	^{255}Lu	0.000	30.1
^{261}Hg	0.000	32.1	^{257}Ta	0.002	29.6
^{256}Hf	0.000	28.9	^{254}Re	0.087	29.2
^{258}W	0.000	28.3	^{256}Ta	0.031	28.6
^{253}W	0.086	28.2	^{259}Re	0.003	28.5

The density distributions of ^{263}Hg, ^{256}Hf, ^{255}Lu, and ^{254}Re are presented in Figure 1, representing even–odd, even–even, odd–even, and odd–odd nuclei, respectively. The scaled density distributions of protons $\rho_p A/Z$ and neutrons $\rho_n A/N$ are provided to facilitate comparison with the total baryon density. Since nuclear interactions depend on the nucleon composition, direct comparisons between different nuclides are challenging due to variations in the mass numbers, which correspond to different nuclear systems. Nevertheless, a qualitative analysis provides useful insights. First, both ^{255}Lu and ^{256}Hf have a neutron magic number of 184. Compared to the other two nuclides, their neutron density distributions appear relatively smooth. This suggests that deviation from a neutron magic number significantly affects the overall nuclear density distribution. Furthermore, an analysis of proton density distributions indicates that both odd and even proton numbers consistently result in lower central densities. However, inherent differences between these nuclear systems make direct quantitative comparisons challenging.

Figure 2 presents the proton single-particle levels for the ground states of ^{263}Hg, ^{256}Hf, ^{255}Lu, and ^{254}Re. The states above the Fermi surface appear due to the inclusion of the pairing potential in the DRHBc framework. For the spherical nuclei ^{263}Hg, ^{256}Hf, and ^{255}Lu, only the $l = 0$ orbital contributes to the central density because angular momentum l is a conserved quantum number. In contrast, ^{254}Re exhibits slight quadrupole deformation ($\beta_2 = 0.087$), leading to the splitting of single-particle states. Despite this deformation, a low occupation probability of 3s orbital is observed in all nuclei, resulting in a depleted central proton density. For even-Z nuclei, the 1s and 2s states are fully occupied because their energy levels lie far below the Fermi surface, consistent with previous findings [3]. In odd-Z nuclei, however, one proton occupies either the $11/2^-$ or $9/2^-$ state near the Fermi surface with an occupation probability of 0.5, rather than the $1/2^+$ state. These high-angular-momentum states ($l > 9/2$) contribute minimally to the central density due to their spatial distribution, effectively acting as blocked protons. This suggests that a lower central density reduces destabilizing interactions between the surface and center, leading to a more stable nuclear structure. In other words, structures with lower central proton densities tend to be more stable than those with denser centers.

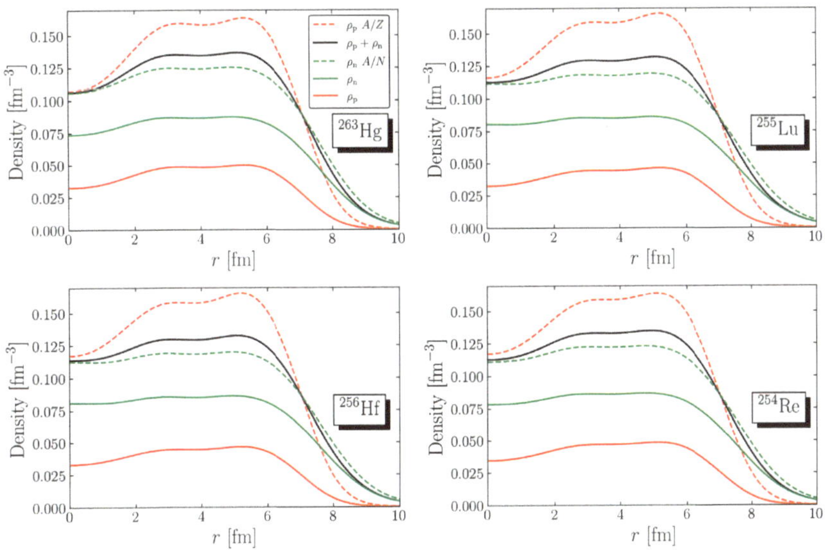

Figure 1. Density profiles for ^{263}Hg, ^{255}Lu, ^{256}Hf, ^{254}Re. For comparison with the total baryon density, the scaled proton density $\rho_p\ A/Z$ and neutron density $\rho_n\ A/N$ are also displayed.

In the case of the deformed nucleus ^{254}Re, the angular momentum l is not a good quantum number, but its projection onto the symmetry axis remains well defined. Consequently, the $1/2^+$ states in deformed nuclei can still contribute to the central density. The occupancy of the s state in deformed nuclei is estimated using the formalism from [24]:

$$N_{nlj}^{\mathrm{DRHBc}} = \langle \Psi | \hat{N}_{nlj} | \Psi \rangle = \langle \Psi | \sum_m c_{njlm}^{\dagger} c_{njlm} | \Psi \rangle, \tag{6}$$

where m stands for the total angular momentum projection on the symmetry axis. For ^{254}Re, as shown in Figure 2, the occupation of 3s states remains low, leading to a depleted central region. Both ^{255}Lu, and ^{254}Re are odd-Z nuclides, exhibiting the odd-proton blocking effect. As previously mentioned, ^{254}Re features more single-particle levels in its ground state than ^{263}Hg, ^{256}Hf, and ^{255}Lu. This is attributed to quadrupole deformation, which alters the nuclear structure and enables the inclusion of the $1/2^+s$ state. These results suggest that in odd-Z nuclides, nuclear deformation influences the bubble structure but does not necessarily suppress it, instead modifying the spatial distribution of nucleons within the nucleus.

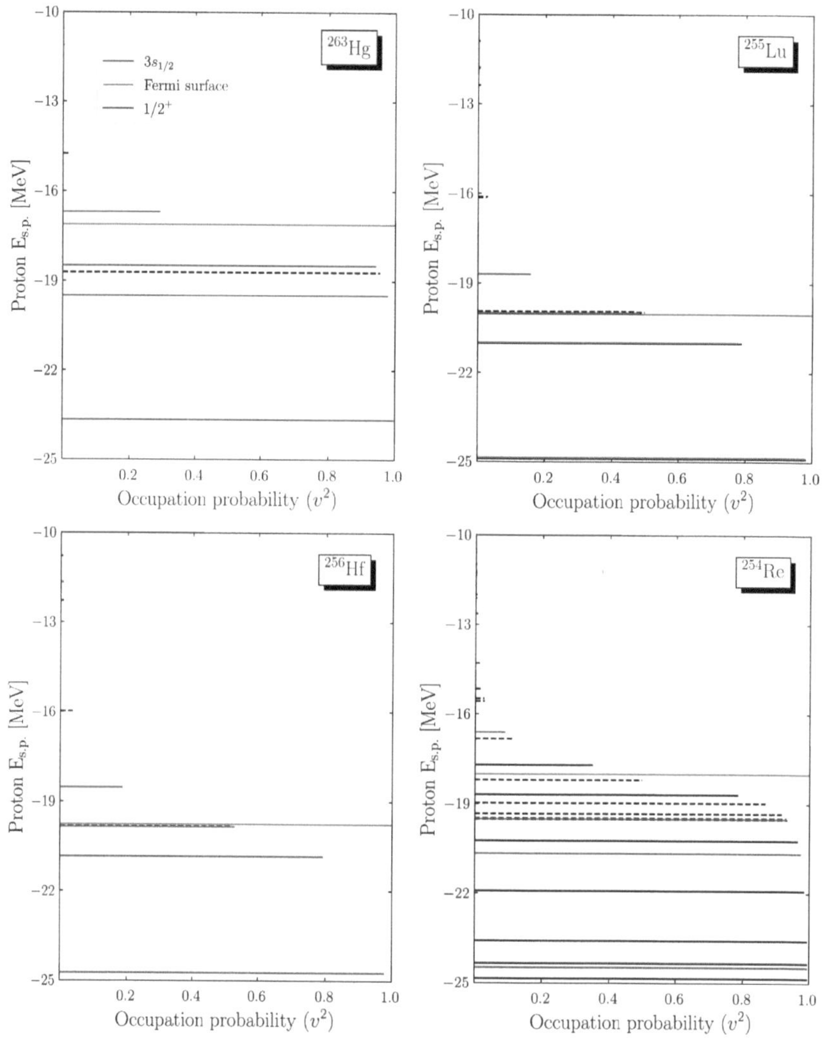

Figure 2. Proton single-particle levels near the Fermi surface for ^{263}Hg, ^{255}Lu, ^{256}Hf, and ^{254}Re. The solid line means the positive-parity state and dashed line means the negative parity. Only $1/2^+$ states contribute to the *s* states. Note that the occupation probability of the 3*s* state of the near Fermi surface is low.

4. Conclusions

In this study, we systematically calculated the bubble parameter for nuclei with $71 \leq Z \leq 80$, extending the scope of the previous research [3]. We first identified the five nuclei with the highest bubble parameter in both even-*Z* and odd-*Z* nuclei groups. These ten nuclei were found to be either spherical or only slightly deformed, consistent with previous findings that nuclear deformation weakens the bubble structure. Next, we selected ^{263}Hg, ^{256}Hf, ^{255}Lu, and ^{254}Re as representative examples of even–odd, even–even, odd–even, and odd–odd bubble nuclei. These nuclei were analyzed in detail through their density distributions and proton single-particle levels. As anticipated from previous studies, the depletion in central density is primarily driven by lower occupations of *s* states near the Fermi surface. Furthermore, our results emphasized that that quadrupole deformation

(β_2) plays a critical role in moderating the bubble structure, with more deformed nuclei exhibiting weaker central density depletion. Additionally, for odd-Z nuclei such as ^{255}Lu and ^{254}Re, the proton blocking effect was observed, influencing the distribution of single-particle levels and contributing to the distinct characteristics of central density depletion in these nuclei. These findings provide further insight into the interplay between deformation, single-particle structure, and central density depletion, advancing our understanding of bubble structures in exotic nuclei.

Author Contributions: Conceptualization, Y.K.; methodology, Y.C.; formal analysis, C.S. and Y.C.; investigation, C.S. and Y.C.; writing—original draft preparation, C.S. and Y.C.; writing—review and editing, All authors; visualization, C.S. and Y.C.; supervision, C.-H.L.; project administration, C.-H.L.; funding acquisition, C.-H.L. All authors have read and agreed to the published version of the manuscript.

Funding: C.S. and C.-H.L. were supported by National Research Foundation of Korea (NRF) grants funded by the Korean government (Ministry of Science and ICT) (No. RS-2023-NR076639, and No. 2018R1D1A1B07048599). C.S. also was supported in part by Basic Science Research Program through the National Research Foundation of Korea (NRF) funded by the Ministry of Education (Grant Nos. RS-2024-00460031). Y.C. was supported in part by the National Key R&D Program of China (2022YFA1602401) and the National Natural Science Foundation of China (No. 12335009 & 12435010). This work was supported partly by the Rare Isotope Science Project of Institute for Basic Science funded by Ministry of Science, ICT, and Future Planning and NRF (No. 2013M7A1A1075764). A portion of the computational resources were provided by the National Supercomputing Center including technical support (No. KSC-2020-CRE-0329 and No. KSC-2021-CRE-0126).

Data Availability Statement: The data supporting the reported results can be found in the publicly available dataset at the following website: https://drhbctable.jcnp.org/, accessed on 20 March 2025.

Acknowledgments: C.S. and C.-H.L. were supported by National Research Foundation of Korea (NRF) grants funded by the Korean government (Ministry of Science and ICT) (No. RS-2023-NR076639, and No. 2018R1D1A1B07048599). C.S. also was supported in part by Basic Science Research Program through the National Research Foundation of Korea (NRF) funded by the Ministry of Education (Grant Nos. RS-2024-00460031). Y.C. was supported in part by the National Key R&D Program of China (2022YFA1602401) and the National Natural Science Foundation of China (No. 12335009 & 12435010). This work was in part supported by the Institute for Basic Science, Korea (Grant No. IBS-R031-D1).

Conflicts of Interest: The authors declare no conflicts of interest.

References

1. Tarasov, O.B.; Gade, A.; Fukushima, K.; Hausmann, M.; Kwan, E.; Portillo, M.; Smith, M.; Ahn, D.S.; Bazin, D.; Chyzh, R.; et al. Observation of New Isotopes in the Fragmentation of Pt198 at FRIB. *Phys. Rev. Lett.* **2024**, *132*, 072501. [CrossRef] [PubMed]
2. Dronchi, N.; Charity, R.J.; Sobotka, L.G.; Brown, B.A.; Weisshaar, D.; Gade, A.; Brown, K.W.; Reviol, W.; Bazin, D.; Farris, P.J.; et al. Evolution of shell gaps in the neutron-poor calcium region from invariant-mass spectroscopy of Sc37,38, Ca35, and K34. *Phys. Rev. C* **2024**, *110*, L031302. [CrossRef]
3. Choi, Y.B.; Lee, C.H.; Mun, M.H.; Kim, Y. Bubble nuclei with shape coexistence in even-even isotopes of Hf to Hg. *Phys. Rev. C* **2022**, *105*, 024306. [CrossRef]
4. Wong, C. Bubble nuclei. *Phys. Lett. B* **1972**, *41*, 451–454. [CrossRef]
5. Todd-Rutel, B.G.; Piekarewicz, J.; Cottle, P.D. Spin orbit splitting in low j neutron orbits and proton densities in the nuclear interior. *Phys. Rev. C* **2004**, *69*, 021301. [CrossRef]
6. Yao, J.M.; Baroni, S.; Bender, M.; Heenen, P.H. Beyond-mean-field study of the possible 'bubble' structure of 34Si. *Phys. Rev. C* **2012**, *86*, 014310. [CrossRef]
7. Wang, Y.; Hou, Z.; Zhang, Q.; Tian, R.; Gu, J. Effect of a tensor force on the proton bubble structure of Hg206. *Phys. Rev. C* **2015**, *91*, 017302. [CrossRef]
8. Duguet, T.; Somà, V.; Lecluse, S.; Barbieri, C.; Navrátil, P. Ab initio calculation of the potential bubble nucleus ^{34}Si. *Phys. Rev. C* **2017**, *95*, 034319. [CrossRef]

9. Saxena, G.; Kumawat, M.; Kaushik, M.; Jain, S.; Aggarwal, M. Bubble Structure in Magic Nuclei. *Phys. Lett. B* **2019**, *788*, 1–6. [CrossRef]
10. Li, J.J.; Long, W.H.; Margueron, J.; Van Giai, N. ^{48}Si: An atypical nucleus? *Phys. Lett. B* **2019**, *788*, 192–197. [CrossRef]
11. Arthuis, P.; Barbieri, C.; Vorabbi, M.; Finelli, P. *AbInitio* Computation of Charge Densities for Sn and Xe Isotopes. *Phys. Rev. Lett.* **2020**, *125*, 182501. [CrossRef] [PubMed]
12. Kumawat, M.; Saxena, G.; Kaushik, M.; Jain, S.; Deegwal, J.; Aggarwal, M. Novel feature of doubly bubble nuclei in $50 \leq Z(N) \leq 82$ region along with magicity and weakly bound structure. *Int. J. Mod. Phys. E* **2020**, *29*, 2050068. [CrossRef]
13. Khan, E.; Grasso, M.; Margueron, J.; Van Giai, N. Detecting bubbles in exotic nuclei. *Nucl. Phys. A* **2008**, *800*, 37–46. [CrossRef]
14. Grasso, M.; Gaudefroy, L.; Khan, E.; Niksic, T.; Piekarewicz, J.; Sorlin, O.; Giai, N.; Vretenar, D. Nuclear 'bubble' structure in Si-34. *Phys. Rev. C* **2009**, *79*, 034318. [CrossRef]
15. Shukla, A.; Åberg, S. Deformed bubble nuclei in the light-mass region. *Phys. Rev. C* **2014**, *89*, 014329. [CrossRef]
16. Luo, Z.J.; Yu, K.; Zhou, X.R.; Cui, J.W.; Sagawa, H. Effects of deformation, pairing and tensor correlation on the evolution of bubble structure within the Skyrme-Hartree-Fock method. *Eur. Phys. J. A* **2018**, *54*, 193. [CrossRef]
17. Saxena, G.; Kumawat, M.; Agrawal, B.; Aggarwal, M. A systematic study of the factors affecting central depletion in nuclei. *J. Phys. G* **2019**, *46*, 065105. [CrossRef]
18. Meng, J.; Toki, H.; Zhou, S.G.; Zhang, S.Q.; Long, W.H.; Geng, L.S. Relativistic Continuum Hartree Bogoliubov theory for ground state properties of exotic nuclei. *Prog. Part. Nucl. Phys.* **2006**, *57*, 470–563. [CrossRef]
19. Zhang, K.; Cheoun, M.-K.; Choi, Y.-B.; Chong, P.S.; Dong, J.; Geng, L.; Ha, E.; He, X.; Heo, C.; Ho, M.C.; et al. Deformed relativistic Hartree-Bogoliubov theory in continuum with point coupling functional: Examples of even-even Nd isotopes. *Phys. Rev. C* **2020**, *102*, 024314. [CrossRef]
20. Pan, C.; Cheoun, M.-K.; Choi, Y.-B.; Dong, J.; Du, X.; Fan, X.H.; Gao, W.; Geng, L.; Ha, E.; He, X.-T.; et al. Deformed relativistic Hartree-Bogoliubov theory in continuum with a point-coupling functional. II. Examples of odd Nd isotopes. *Phys. Rev. C* **2022**, *106*, 014316. [CrossRef]
21. Guo, P.; Cao, X.; Chen, K.; Chen, Z.; Cheoun, M.-K.; Choi, Y.-B.; Lam, P.C.; Deng, W.; Dong, J.; Du, P.; et al. Nuclear mass table in deformed relativistic Hartree–Bogoliubov theory in continuum, II: Even-Z nuclei. *Atom. Data Nucl. Data Tabl.* **2024**, *158*, 101661. [CrossRef]
22. Kucharek, H.; Ring, P. Relativistic field theory of superfluidity in nuclei. *Hadron. Nucl.* **1991**, *339*, 23–35. [CrossRef]
23. Zhao, P.; Li, Z.; Yao, J.; Meng, J. New parametrization for the nuclear covariant energy density functional with point-coupling interaction. *Phys. Rev. C* **2010**, *82*, 054319. [CrossRef]
24. Sun, X.X.; Zhou, S.G. Rotating deformed halo nuclei and shape decoupling effects. *Sci. Bull.* **2021**, *66*, 2072–2078. [CrossRef]

particles

MDPI

Article

Calculation of α Decay Half-Lives for Tl, Bi, and At Isotopes

Myeong-Hwan Mun [1,2,*], Kyoungsu Heo [2] and Myung-Ki Cheoun [2]

1 Department of Physics, Kyungpook National University, Daegu 41566, Republic of Korea
2 Department of Physics and Origin of Matter and Evolution of Galaxies (OMEG) Institute, Soongsil University, Seoul 06978, Republic of Korea; pleasewhy@naver.com (K.H.); cheoun@ssu.ac.kr (M.-K.C.)
* Correspondence: aa3101@gmail.com

Abstract: We investigated the reaction Q-value (Q_α) for the α decay of Tl, Bi, and At isotopes using the deformed relativistic Hartree–Bogoliubov theory in continuum (DRHBc) with the covariant density functional PC-PK1. The α decay half-lives of Tl, Bi, and At isotopes are evaluated using various empirical formulas, based on both experimental Q_α and those obtained from DRHBc calculations. The calculated Q_α and α decay half-lives are compared with experimental data. On the basis of these results, we also predicted the α decay half-lives of isotopes for which experimental data are unavailable.

Keywords: deformed relativistic Hartree–Bogoliubov theory in continuum (DRHBc); α decay half-lives; empirical formula

1. Introduction

Various nuclear decay modes have been experimentally observed, including α decay, β decay, proton decay, neutron decay, spontaneous fission, and electron capture [1]. Among these, α decay is one of the most crucial decay modes, widely recognized as a key tool for investigating unstable and neutron-deficient isotopes, as well as superheavy elements. Since its discovery by Becquerel in 1896, α decay has become a major research topic in nuclear physics. This decay mode also provides essential insights into the structure and stability of atomic nuclei, as well as the mechanisms behind decay. Additionally, α decay plays a crucial role in the synthesis of superheavy elements, providing valuable insights for predicting and understanding the existence of specific elements. For example, it enables the prediction of the half-lives of superheavy elements and the discovery of new decay pathways. As such, α decay remains a central focus of various nuclear physics research.

Owing to the significant advancements in experimental technology, substantial progress has been made in both the experimental [2,3] and theoretical [4,5] aspects of α decay. Experimentally, various heavy nuclei have been successfully discovered through the analysis of alpha decay chains in recent years. For example, ^{214}U, a new α-emitting nucleus, has been successfully produced through the ^{182}W(^{36}Ar, 4n)^{214}U reaction [2]. Theoretically, several empirical formulas have been developed to study α decay half-lives, including the Royer formula [6], AKRA [7], Viola–Seaborg–Sobiczewski (VSS) formula [8,9], Sobiczewski–Parkhomenko (SP) formula [10], Universal Decay Law (UDL) [11,12], and others.

Q_α is one of the significant characteristic quantities of an alpha-emitting nucleus. It is given as follows:

$$Q_\alpha = E_b(Z-2, N-2) - E_b(Z, N) + E_b(2,2), \tag{1}$$

where E_b is the binding energy of the nucleus, and the binding energy of ^{4}He (E_b(2,2)) is 28.30 MeV.

The nuclear masses of over 2000 nuclei have been experimentally measured. However, α decay is still anticipated to occur in the vast, unexplored regions of the nuclear chart, which remain beyond the reach of experimental techniques in the near future. Therefore, a detailed analysis of Q_α must depend on reliable theoretical nuclear mass models.

For addressing the mentioned issue, a well-refined and state-of-the-art relativistic nuclear model is essential. This model should simultaneously account for the deformation, pairing correlations, and continuum effects within a microscopic framework capable of covering the entire nuclear mass range. In this context, the deformed relativistic Hartree–Bogoliubov theory in continuum (DRHBc), based on point-coupling density functionals, has been developed [13,14]. The DRHBc theory has been shown to provide a robust description of nuclear masses with high predictive power [15,16], and it has also been applied to study the nuclear structure of various isotopes [17–25].

Generally, even–even nuclei are more suitable for α decay studies due to their abundant experimental data, well-defined decay paths, and theoretical simplicity. In Ref. [26], the α decay of even–even actinides and super-heavy nuclei with $Z \geq 90$ was carefully analyzed. Additionally, in our previous study [27], we investigated the α-decay half-lives of even–even nuclei in the range from W to U using DRHBc theory and the semiclassical WKB approximation. In this article, we investigate the α decay half-lives of Tl, Bi, and At isotopes using empirical formulas and the DRHBc theory with the PC-PK1 density functional [28]. The article is organized as follows. Section 2 introduces a brief overview of the DRHBc theory and the empirical formulas used in this study, along with the numerical details for DRHBc calculations. The results and discussions for Tl, Bi, and At isotopes are presented in Section 3. Finally, the summary and conclusions are provided in Section 4.

2. Theoretical Framework

2.1. Deformed Relativistic Hartree–Bogoliubov Theory in Continuum

The detailed formalism of the DRHBc theory can be found in Refs. [29–31]. Here, we provide only a brief overview of the formalism of the DRHBc theory. In the DRHBc theory, the relativistic Hartree–Bogoliubov (RHB) equation [32] is expressed as follows.

$$\begin{pmatrix} h_D - \lambda & \Delta \\ -\Delta^* & -h_D^* + \lambda \end{pmatrix} \begin{pmatrix} U_k \\ V_k \end{pmatrix} = E_k \begin{pmatrix} U_k \\ V_k \end{pmatrix}. \tag{2}$$

Here, λ is the Fermi energy, and E_k and $(U_k, V_k)^T$ are the quasiparticle energy and quasiparticle wave function. In the coordinate space, the Dirac Hamiltonian h_D can be defined as

$$h_D(r) = \alpha \cdot p + V(r) + \beta[M + S(r)], \tag{3}$$

where α and p are the Dirac matrices and the momentum operator, M is the nucleon mass, and $V(r)$ and $S(r)$ are the vector and scalar potentials, respectively. The paring potential Δ is expressed in terms of the pairing tensor $\kappa(r, r')$ as follows:

$$\Delta(r, r') = V(r, r')\kappa(r, r'), \tag{4}$$

using a density-dependent zero range force

$$V(r, r') = \frac{V_0}{2}(1 - P^\sigma)\delta(r - r')\left(1 - \frac{\rho(r)}{\rho_{sat}}\right). \tag{5}$$

Here, P^σ is the spin exchange operator, and ρ_{sat} is the nuclear saturation density.

The total energy can be computed as

$$
\begin{aligned}
E_{\text{tot}} &= \sum_{k>0}(\lambda_\tau - E_k)v_k^2 - \frac{1}{2}\int d^3r\kappa(r)\Delta(r) + E_{\text{c.m.}} \\
&\quad - \int d^3r\Big(\frac{1}{2}\alpha_S\rho_S^2 + \frac{1}{2}\alpha_V\rho_V^2 + \frac{1}{2}\alpha_{TV}\rho_{TV}^2 + \frac{2}{3}\beta_S\rho_S^3 + \frac{3}{4}\gamma_S\rho_S^4 + \frac{3}{4}\gamma_V\rho_V^4 \\
&\quad + \frac{1}{2}\delta_S\rho_S\Delta\rho_S + \frac{1}{2}\delta_V\rho_V\Delta\rho_V + \frac{1}{2}\delta_{TV}\rho_{TV}\Delta\rho_{TV} + \frac{1}{2}\rho_p eA^0\Big),
\end{aligned}
\tag{6}
$$

where E_k is the single-particle energy, v_k^2 is the occupation probability, e is the unit of charge, and A^0 is the electromagnetic field, respectively. The coupling constant α_i for the four-fermion terms is specified by the superscripts ($i = S, V$ and T), which stand for the scalar, vector, and isovector channels, respectively. β_S, γ_S, and γ_V are the higher-order terms, while δ_i refers to the gradient terms. Finally, $E_{\text{c.m.}}$ denotes the center of mass correction energy.

For the numerical calculations of the Tl, Bi, and At isotopes, we employ the energy cut-off $E^+_{cut} = 300$ MeV and the angular momentum cutoff $J_{max} = (23/2)\hbar$ for the Dirac Woods–Saxon basis. The pairing strength $V_0 = -325.0$ MeV fm^3, a pairing window of 100 MeV, and a saturation density of $\rho_{sat} = 0.152$ fm^{-3} are taken, respectively. The numerical details can be found in Refs. [13,14].

2.2. Empirical Formula for α Decay Half-Lives

Empirical formulas for the α decay half-lives typically depend on the proton number (Z), the mass number (A), and the reaction Q-value (Q_α) for the α decay. The most crucial factor in the α decay process of the heavy nuclei is the accurate determination of Q_α, as it reflects the structure of the heavy nuclei through the binding energy. The significance of Q_α is clearly mentioned in Refs. [8,33]. For Q_α, we use both experimental data and DRHBc mass table data, particularly when experimental Q_α are unavailable.

2.2.1. Royer Formula

The Royer formula [6] is given by

$$\log_{10} T_{1/2} = a + bA^{1/6}\sqrt{Z} + \frac{cZ}{\sqrt{Q_\alpha}}, \tag{7}$$

where A, Z, and Q_α are the mass number, proton number, and the reaction Q-value for the α decay, respectively. Additionally, the parameters a, b, and c are $a = -25.68$, $b = -1.1423$, and $c = 1.592$ for odd Z–even N nuclei, and $a = -29.48$, $b = -1.113$, and $c = 1.6971$ for odd Z–odd N nuclei, respectively [34].

2.2.2. AKRA Formula

Akrawy and Poenaru presented [7] a new expression for calculating the α decay half-life by incorporating nuclear isospin asymmetry $I = (N - Z)/A$.

$$\log_{10} T_{1/2} = a + bA^{1/6}\sqrt{Z} + \frac{cZ}{\sqrt{Q_\alpha}} + dI + eI^2. \tag{8}$$

For the odd Z–even N nuclei case (odd Z–odd N nuclei case), the parameters a, b, c, d, and e are as follows: $a = -31.79248$ (-26.27896), $b = -1.07636$ (-1.20130), $c = 1.75354$ (1.65906), $d = -2.22627$ (-0.08411), and $e = 0.39378$ (67.59728) [35].

2.2.3. Viola–Seaborg–Sobiczewski (VSS) Formula

The VSS [8,9] formula proposed by Sobiczewski extended the original Viola–Seaborg formula to better account for heavy and superheavy nuclei are widely used to calculate and predict the α decay half-lives.

$$\log_{10} T_{1/2} = \frac{aZ + b}{\sqrt{Q_\alpha}} + cZ + d + h_{log}, \tag{9}$$

where $a = 1.66175$, $b = -8.5166$, $c = -0.20228$, and $d = -33.9069$, respectively. The term h_{log} describes the hindrance effects related to odd-Z and/or odd-N. Its value is 0.772 for odd-Z even-N nuclei and 1.114 for odd-Z odd-N nuclei [36].

2.2.4. Parkhomenko–Sobiczewski (SP) Formula

The Parkhomenko–Sobiczewski (SP) formula, which is a phenomenological expression used to explain the α decay half-lives of nuclei heavier than ^{208}Pb, was introduced by Parkhomenko and Sobiczewski [10]. The (SP) formula is given by

$$\log_{10} T_{1/2} = \frac{aZ}{\sqrt{Q_\alpha - E_i}} + bZ + c, \tag{10}$$

where the values of the coefficients [36] are $a = 1.5372$, $b = -0.1607$, and $c = -36.573$. The E_i represents the average excitation energy, with values of 0.113 and 0.284 for odd–even and odd–odd nuclei, respectively.

2.2.5. Universal Decay Law (UDL) Formula

Qi et al. [11,12] derived a linear universal decay law (UDL) based on an R-matrix theory that describes the microscopic mechanism of α emission and is applicable to α decay. The UDL formula is expressed as follows:

$$\log_{10} T_{1/2} = aZ_pZ_d\sqrt{\frac{\mu}{Q_\alpha}} + b\sqrt{\mu Z_pZ_d(A_p^{1/3} + A_d^{1/3})} + c. \tag{11}$$

Here, $\mu = A_aA_d/(A_a + A_d)$, where A_a denotes the mass number of the emitted alpha particle and A_d represents the mass number of the daughter nucleus. In Equation (11), the coefficients for the UDL formula, as provided in Ref. [35], are as follows: $a = 0.4314$, $b = -0.4087$, and $c = -25.7725$.

3. Results

It is well-known that the α decay half-lives are highly sensitive to Q_α. Therefore, selecting an accurate Q_α is crucial for making reliable predictions. First, we examined the Q_α of the Tl, Bi, and At by the DRHBc theory. In Figure 1, the Q_α for Tl, Bi, and At obtained from the DRHBc calculations are plotted against the neutron number, along with the available experimental data [37]. Additionally, for a quantitative comparison, we present the differences between the calculated results and the experimental data, with uncertainties represented by standard deviation of less than 1.14, 1.03, and 0.95 MeV for Tl, Bi, and At, respectively, as shown in Figure 1.

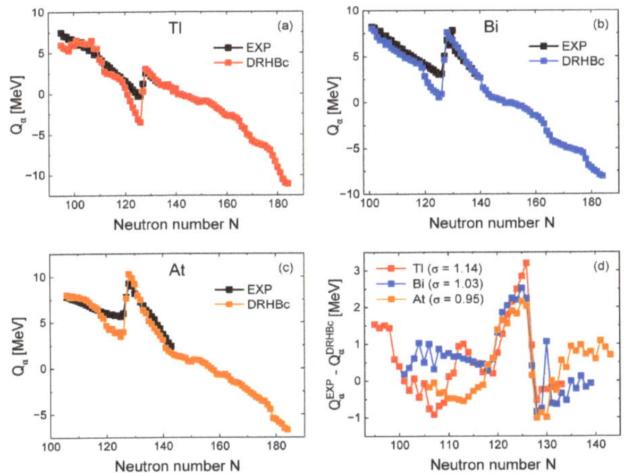

Figure 1. Q_α for (**a**) Tl, (**b**) Bi, and (**c**) At isotopes determined by DRHBc calculations. (**d**) The difference between the calculated results and the experimental data. They are compared with available experimental data taken from Ref. [37]. The numbers in parentheses stand for standard deviation in MeV to the data [37].

The α decay half-lives calculated using five empirical formulas (AKRA, Royer, SP, UDL, and VSS) are presented in Figure 2. The results were obtained using the experimental Q_α^{EXP} values and the Q_α^{DRHBc} values derived from DRHBc calculations, as shown in panels (a) and (b) of Figure 2, respectively. For comparison with the experiment, we use the experimental α decay half-lives from NNDC [38]. Since multiple decay modes can exist for each nucleus, we consider only 17 experimental data points where the branching ratio of α decay is close to 100%. The logarithmic differences between the experimental half-lives and the calculated values are shown in Figure 2c,d. Additionally, the calculated α decay half-lives are listed in Tables 1 and 2. Table 1 shows the results derived using experimental Q_α^{EXP} values, while Table 2 displays the results obtained using Q_α^{DRHBc} values from DRHBc calculations. The standard deviations σ between the experimental data and the calculated results, as defined by

$$\sigma = \sqrt{\frac{1}{N}\Sigma(log_{10}T_{1/2}^{EXP} - log_{10}T_{1/2}^{cal})^2}, \tag{12}$$

are provided in the last row of Tables 1 and 2. The Q_α calculated based on the DRHBc masses has uncertainty due to the deviation from experimental values (σ = 2.2 MeV). This uncertainty can also affect the accuracy of the α decay half-life calculated using Q_α^{DRHBc}. The results are summarized in Table 2.

The predicted α decay half-lives of Tl (Z = 81), Bi (Z = 83), and At (Z = 85) isotopes, calculated using the VSS and SP formulas—two models with the smallest standard deviations among the five models, as shown in Table 2—are shown in Figure 3a,b, with Q_α values taken from AME2020 [37] and the DRHBc mass table. We also compared the results obtained using Q_α values derived from AME2020 and the DRHBc mass table. For a given isotope chain, the calculated half-lives generally increase as the neutron number increases. This trend decreases near the N = 126 shell closure and then increases again.

Table 1. The calculated half-life of α decay using selected empirical formulas. The experimental data for Q_α^{EXP} and half-lives are taken from AME2020 [37] and NNDC [38], respectively. The units of Q_α^{EXP} and $\log_{10} T_{1/2}$ are MeV and seconds (s), respectively.

α Transition	Q_α^{EXP}	$\log_{10} T_{1/2}$					
		EXP	AKRA	Royer	SP	UDL	VSS
$^{184}Bi \rightarrow {}^{180}Tl$	8.22	−1.89	−3.71	−4.53	−3.56	−5.23	−4.45
$^{186}Bi \rightarrow {}^{182}Tl$	7.76	−1.83	−2.21	−3.13	−2.15	−3.85	−3.12
$^{187}Bi \rightarrow {}^{183}Tl$	7.76	−1.43	−3.23	−3.12	−2.68	−3.87	−3.46
$^{188}Bi \rightarrow {}^{184}Tl$	7.26	−1.22	−0.47	−1.49	−0.49	−2.23	−1.57
$^{189}Bi \rightarrow {}^{185}Tl$	7.27	−0.16	−1.56	−1.60	−1.10	−2.26	−1.92
$^{191}At \rightarrow {}^{187}Bi$	7.82	−2.77	−2.55	−2.57	−2.07	−3.27	−2.87
$^{192}At \rightarrow {}^{188}Bi$	7.70	−1.94	−1.17	−2.13	−1.12	−2.87	−2.14
$^{193}At \rightarrow {}^{189}Bi$	7.57	−1.55	−1.74	−1.82	−1.27	−2.47	−2.09
$^{194}At \rightarrow {}^{190}Bi$	7.45	−0.54	−0.25	−1.33	−0.29	−2.09	−1.37
$^{195}At \rightarrow {}^{191}Bi$	7.34	−0.51	−0.97	−1.11	−0.50	−1.72	−1.35
$^{212}At \rightarrow {}^{208}Bi$	7.82	−0.50	−0.25	−2.94	−1.51	−3.62	−2.51
$^{213}At \rightarrow {}^{209}Bi$	9.25	−6.90	−7.48	−6.93	−6.00	−7.75	−6.70
$^{214}At \rightarrow {}^{210}Bi$	8.99	−6.25	−3.49	−6.46	−4.91	−7.08	−5.71
$^{215}At \rightarrow {}^{211}Bi$	8.18	−4.43	−4.41	−4.14	−3.15	−4.81	−3.91
$^{216}At \rightarrow {}^{212}Bi$	7.95	−3.52	−0.35	−3.45	−1.94	−4.12	−2.91
$^{217}At \rightarrow {}^{213}Bi$	7.20	−1.49	−1.04	−1.07	−0.01	−1.58	−0.87
$^{218}At \rightarrow {}^{214}Bi$	6.88	0.11	3.59	0.36	1.85	−0.36	0.63
standard deviation			1.55	1.11	1.09	1.61	1.16

Table 2. The same as Table 1, but with results obtained using Q_α^{DRHBc} from the DRHBc calculations. The standard deviation ($\pm\sigma$) are the results obtained using Q_α^{DRHBc} ($\pm\sigma$) ($\sigma = 2.2$), respectively.

α Transition	Q_α^{DRHBc}	$\log_{10} T_{1/2}$					
		EXP	AKRA	Royer	SP	UDL	VSS
$^{184}Bi \rightarrow {}^{180}Tl$	8.04	−1.89	−3.18	−3.98	−3.02	−4.69	−3.94
$^{186}Bi \rightarrow {}^{182}Tl$	7.15	−1.83	−0.16	−1.03	−0.08	−1.79	−1.19
$^{187}Bi \rightarrow {}^{183}Tl$	6.75	−1.43	0.52	0.28	0.76	−0.31	−0.13
$^{188}Bi \rightarrow {}^{184}Tl$	6.77	−1.22	1.36	0.38	1.36	−0.39	0.15
$^{189}Bi \rightarrow {}^{185}Tl$	6.27	−0.16	2.57	2.15	2.70	1.66	1.75
$^{191}At \rightarrow {}^{187}Bi$	8.00	−2.77	−3.15	−3.11	−2.62	−3.84	−3.40
$^{192}At \rightarrow {}^{188}Bi$	7.76	−1.94	−1.37	−2.33	−1.32	−3.08	−2.33
$^{193}At \rightarrow {}^{189}Bi$	7.92	−1.55	−2.94	−2.91	−2.37	−3.62	−3.16
$^{194}At \rightarrow {}^{190}Bi$	7.77	−0.54	−1.30	−2.41	−1.35	−3.14	−2.36
$^{195}At \rightarrow {}^{191}Bi$	7.83	−0.51	−2.69	−2.67	−2.08	−3.35	−2.88
$^{212}At \rightarrow {}^{208}Bi$	7.67	−0.50	0.22	−2.46	−1.04	−3.15	−2.07
$^{213}At \rightarrow {}^{209}Bi$	10.27	−6.90	−9.97	−9.19	−8.27	−10.11	−8.91
$^{214}At \rightarrow {}^{210}Bi$	9.86	−6.25	−5.61	−8.63	−7.01	−9.21	−7.71
$^{215}At \rightarrow {}^{211}Bi$	9.18	−4.43	−7.34	−6.79	−5.82	−7.59	−6.52
$^{216}At \rightarrow {}^{212}Bi$	7.96	−3.52	−0.37	−3.48	−1.96	−4.14	−2.93
$^{217}At \rightarrow {}^{213}Bi$	7.46	−1.49	−2.03	−1.97	−0.91	−2.52	−1.75
$^{218}At \rightarrow {}^{214}Bi$	6.33	0.11	5.86	2.68	4.15	1.93	2.77
standard deviation			2.33	1.77	1.75	2.11	1.62
standard deviation ($+\sigma$)			6.07	6.62	5.59	7.38	6.24
standard deviation ($-\sigma$)			11.12	9.81	10.98	9.29	9.26

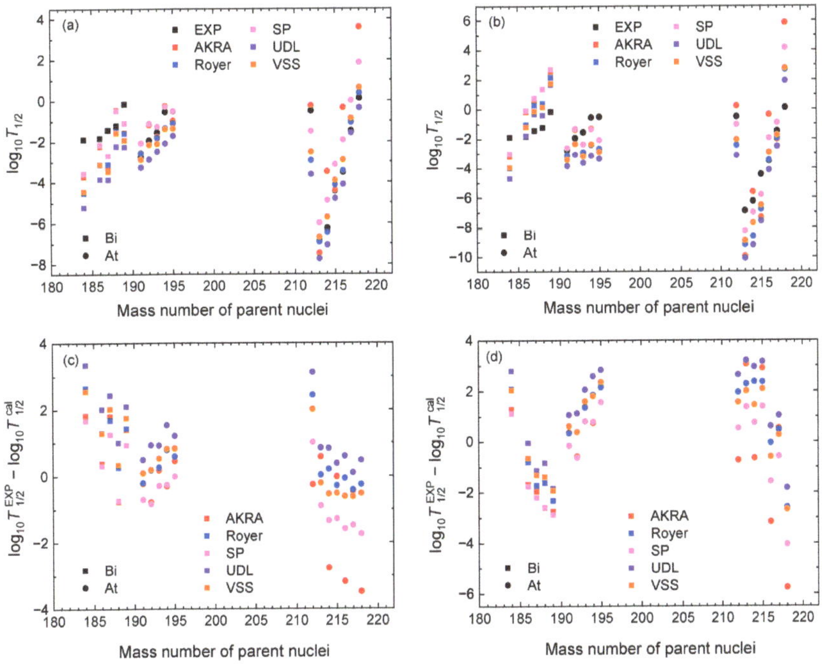

Figure 2. The α decay half-lives obtained of Bi and At isotopes by different five empirical formulas (AKRA, Royer, SP, UDL, and VSS) using (**a**) the experimental Q_{α}^{EXP} values and (**b**) the Q_{α}^{DRHBc} values derived from DRHBc calculations. (**c**,**d**) The logarithmic differences between the experimental half-lives and the calculated values.

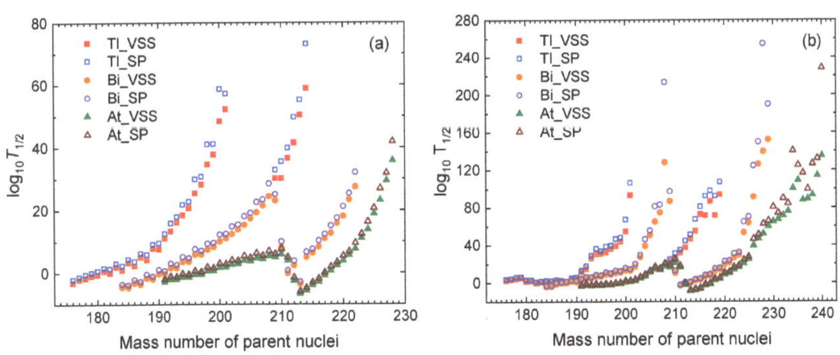

Figure 3. The predicted α decay half-lives in logarithmic form for Tl (Z = 81), Bi (Z = 83), and At (Z = 85) isotopes using the VSS and SP formula with (**a**) available experimental Q_{α}^{EXP} and (**b**) the Q_{α}^{DRHBc} obtained from DRHBc calculations.

In Figure 4a, the logarithmic differences of α decay half-lives for Tl (Z = 81), Bi (Z = 83), and At (Z = 85) isotopes between the results obtained using Q_{α} values derived from AME2020 and those derived from the DRHBc mass table versus the mass number of the parent nucleus are presented. The results of the two formulas, VSS and SP, are nearly identical, except for ^{208}Bi. Additionally, the differences with respect to ($Q_{\alpha}^{EXP} - Q_{\alpha}^{DRHBc}$) are presented in Figure 4b. We can observe that, as the value of ($Q_{\alpha}^{EXP} - Q_{\alpha}^{DRHBc}$) increases, the logarithmic differences between the results predicted using Q_{α} values from AME2020 and those from the DRHBc mass table also increase, particularly when it exceeds 2.0. Finally, we can see that the difference between the results of VSS and SP formulas increases. We will calculate and predict the α decay half-lives using the predicted densities in the DRHBc theory within the WKB approximation framework [39,40] in the following study.

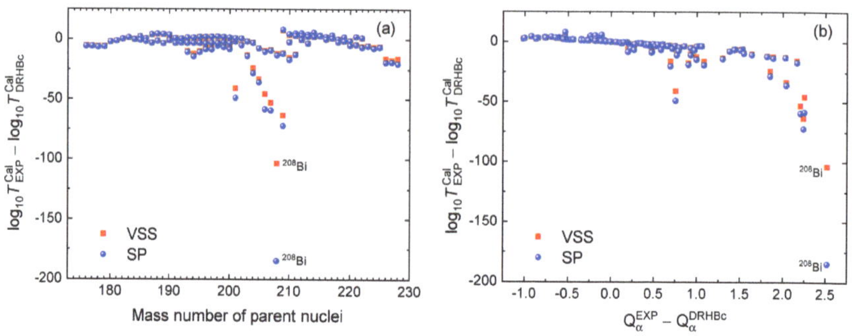

Figure 4. The logarithmic differences of α decay half-lives for Tl (Z = 81), Bi (Z = 83), and At (Z = 85) isotopes, obtained using Q_α values derived from AME2020 and using Q_α^{DRHBc}. (**a**) The DRHBc mass table versus the mass number of the parent nucleus and (**b**) ($Q_\alpha^{EXP} - Q_\alpha^{DRHBc}$).

4. Summary

In this work, we evaluated the reaction Q-value (Q_α) for the α decay of Tl, Bi, and At isotopes using the DRHBc theory and compared the results with experimental data from AME2020. Since multiple decay modes exist for each nucleus, we considered only 17 experimental data points where the branching ratios for α decay modes are almost 100%. The α decay half-lives of these isotopes were calculated using five different empirical formulas, based on both experimental Q_α values and those obtained from the DRHBc calculations. The calculated α decay half-lives were also compared with experimental data from NNDC. The VSS and SP formulas have the smallest standard deviations (σ) between the calculated results and the experimental data among the five models. Based on these results, we calculated and predicted the α decay half-lives of Tl, Bi, and At isotopes using VSS and SP formulas. In the future study, we will calculate and predict the α decay half-lives using the densities obtained from the DRHBc theory within the WKB approximation framework.

Author Contributions: Formal analysis, M.-H.M.; investigation, M.-H.M.; writing—original draft preparation, M.-H.M.; writing—review and editing, all authors. All authors have read and agreed to the published version of the manuscript.

Funding: This work of M.-H.M. was supported by the National Research Foundation of Korea (NRF) grant funded by the Korea government (MSIT) (Grant Nos. NRF-2021R1F1A1060066). M.-K.C. was supported by the NRF grant funded by the Korea government (MSIT) (Grants No. 2020R1A2C3006177 and No. 2021R1A6A1A03043957). K.H. was supported in part by Basic Science Research Program through the National Research Foundation of Korea (NRF) funded by the Ministry of Education (Grant Nos. RS-2024-00460031).

Data Availability Statement: Data will be made available on request.

Acknowledgments: Helpful discussions with members of the DRHBc Mass Table Collaboration are greatly appreciated.

Conflicts of Interest: The authors declare no conflicts of interest.

References

1. Sobiczewski, A.; Pomorski, K. Description of structure and properties of superheavy nuclei. *Prog. Part. Nucl. Phys.* **2007**, *58*, 292.
2. Zhang, Z.Y.; Yang, H.B.; Huang, M.H.; Gan, Z.G.; Yuan, C.X.; Qi, C.; Andreyev, A.N.; Liu, M.L.; Ma, L.; Zhang, M.M.; et al. New α-Emitting Isotope ^{214}U and Abnormal Enhancement of α-Particle Clustering in Lightest Uranium Isotopes. *Phys. Rev. Lett.* **2021**, *126*, 152502.
3. Oganessian, Y.T. Synthesis of the heaviest elements in 48Ca-induced reactions. *Radiochim. Acta* **2011**, *99*, 429.
4. Basu, D.N. Role of effective interaction in nuclear disintegration processes. *Phys. Lett. B* **2003**, *566*, 90.
5. Manjunatha, H.C.; Sowmya, N.; Damodara Gupta, P.S.; Sridhar, K.N.; Nagaraja, A.M.; Seenappa, L.; Alfred Cecil Raj, S. Investigation of decay modes of superheavy nuclei. *Nucl. Sci. Tech.* **2021**, *32*, 130.
6. Royer, G. Alpha emission and spontaneous fission through quasi-molecular shapes. *J. Phys. G* **2000**, *26*, 1149.
7. Akrawy, D.T.; Poenaru, D.N. Alpha decay calculations with a new formula. *J. Phys. G* **2017**, *44*, 105105.
8. Viola, V.E., Jr.; Seaborg, G.T. Nuclear systematics of the heavy elements—II Lifetimes for alpha, beta and spontaneous fission decay. *J. Inorg. Nucl. Chem.* **1966**, *28*, 741.
9. Sobiczewski, A.; Patyk, Z.; Cwiok, S. Deformed superheavy nuclei. *Phys. Lett. B* **1989**, *224*, 1.
10. Parkhomenko, A.; Sobiczewski, A. Phenomenological formula for α-decay half-lives of heaviest nuclei. *Acta Phys. Pol. B* **2005**, *36*, 3095.
11. Qi, C.; Xu, F.R.; Liotta, R.J.; Wyss, R. Universal Decay Law in Charged-Particle Emission and Exotic Cluster Radioactivity. *Phys. Rev. Lett.* **2009**, *103*, 072501. [CrossRef] [PubMed]
12. Qi, C.; Xu, F.R.; Liotta, R.J.; Wyss, R.; Zhang, M.Y.; Asawatangtrakuldee, C.; Hu, D. Microscopic mechanism of charged-particle radioactivity and generalization of the Geiger-Nuttall law. *Phys. Rev. C* **2009**, *80*, 044326. [CrossRef]
13. Zhang, K.; Cheoun, M.K.; Choi, Y.B.; Chong, P.S.; Dong, J.; Geng, L.; Ha, E.; He, X.; Heo, C.; Ho, M.C.; et al. Deformed relativistic Hartree-Bogoliubov theory in continuum with a point-coupling functional: Examples of even-even Nd isotopes. *Phys. Rev. C* **2020**, *102*, 024314. [CrossRef]
14. Pan, C.; Cheoun, M.K.; Choi, Y.B.; Dong, J.; Du, X.; Fan, X.H.; Gao, W.; Geng, L.; Ha, E.; He, X.T.; et al. Deformed relativistic Hartree-Bogoliubov theory in continuum with a point-coupling functional. II. Examples of odd Nd isotopes. *Phys. Rev. C* **2022**, *106*, 014316. [CrossRef]
15. Zhang, K.; Cheoun, M.K.; Choi, Y.B.; Chong, P.S.; Dong, J.; Dong, Z.; Du, X.; Geng, L.; Ha, E.; He, X.T.; et al. Nuclear mass table in deformed relativistic Hartree–Bogoliubov theory in continuum, I: Even–even nuclei. *At. Data Nucl. Data Tables* **2022**, *144*, 101488.
16. Guo, P.; Cao, X.; Chen, K.; Chen, Z.; Cheoun, M.K.; Choi, Y.B.; Lam, P.C.; Deng, W.; Dong, J.; Du, P.; et al. Nuclear mass table in deformed relativistic Hartree–Bogoliubov theory in continuum, II: Even-Z nuclei. *At. Data Nucl. Data Tables* **2024**, *158*, 101661. [CrossRef]
17. Sun, X.-X. Deformed two-neutron halo in ^{19}B. *Phys. Rev. C* **2021**, *103*, 054315. [CrossRef]
18. Yang, Z.H.; Kubota, Y.; Corsi, A.; Yoshida, K.; Sun, X.X.; Li, J.G.; Kimura, M.; Michel, N.; Ogata, K.; Yuan, C.X.; et al. Quasifree Neutron Knockout Reaction Reveals a Small s-Orbital Component in the Borromean Nucleus ^{17}B. *Phys. Rev. Lett.* **2021**, *126*, 082501. [CrossRef]
19. Kim, S.; Mun, M.-H.; Cheoun, M.-K.; Ha, E. Shape coexistence and neutron skin thickness of Pb isotopes by the deformed relativistic Hartree-Bogoliubov theory in continuum. *Phys. Rev. C* **2022**, *105*, 034340. [CrossRef]
20. Zhang, K.Y.; Papakonstantinou, P.; Mun, M.-H.; Kim, Y.; Yan, H.; Sun, X.-X. Collapse of the $N = 28$ shell closure in the newly discovered ^{39}Na nucleus and the development of deformed halos towards the neutron dripline. *Phys. Rev. C* **2023**, *107*, L041303. [CrossRef]
21. Zhang, K.Y.; Yang, S.Q.; An, J.L.; Zhang, S.S.; Papakonstantinou, P.; Mun, M.H.; Kim, Y.; Yan, H. Missed prediction of the neutron halo in ^{37}Mg. *Phys. Lett. B* **2023**, *844*, 138112. [CrossRef]
22. Mun, M.-H.; Kim, S.; Cheoun, M.-K.; So, W.Y.; Choi, S.; Ha, E. Odd-even shape staggering and kink structure of charge radii of Hg isotopes by the deformed relativistic Hartree–Bogoliubov theory in continuum. *Phys. Lett. B* **2023**, *847*, 138298. [CrossRef]
23. Guo, P.; Pan, C.; Zhao, Y.C.; Du, X.K.; Zhang, S.Q. Prolate-shape dominance in atomic nuclei within the deformed relativistic Hartree-Bogoliubov theory in continuum. *Phys. Rev. C* **2023**, *108*, 014319. [CrossRef]
24. An, J.L.; Zhang, K.Y.; Lu, Q.; Zhong, S.Y.; Zhang, S.S. A unified description of the halo nucleus ^{37}Mg from microscopic structure to reaction observables. *Phys. Lett. B* **2024**, *849*, 138422. [CrossRef]
25. Mun, M.-H.; Ha, E.; Sagawa, H.; Colò, G.; Cheoun, M.-K. Symmetry energy from two-nucleon separation energies of Pb and Ca isotopes. *Phys. Rev. C* **2024**, *110*, 014314. [CrossRef]
26. Taninah, A.; Agbemava, S.E.; Afanasjev, A.V. Covariant density functional theory input for *r*-process simulations in actinides and superheavy nuclei: The ground state and fission properties. *Phys. Rev. C* **2020**, *102*, 054330. [CrossRef]
27. Choi, Y.-B.; Lee, C.-H.; Mun, M.-H.; Choi, S.C.; Kim, Y.M. α-decay half-lives for even-even isotopes of W to U. *Phys. Rev. C* **2024**, *109*, 054310. [CrossRef]

28. Zhao, P.W.; Li, Z.P.; Yao, J.M.; Meng, J. New parametrization for the nuclear covariant energy density functional with a point-coupling interaction. *Phys. Rev. C* **2010**, *82*, 054319. [CrossRef]
29. Zhou, S.G.; Meng, J.; Ring, P.; Zhao, E.G. Neutron halo in deformed nuclei. *Phys. Rev. C* **2010**, *82*, 011301. [CrossRef]
30. Li, L.; Meng, J.; Ring, P.; Zhao, E.G.; Zhou, S.G. Deformed relativistic Hartree-Bogoliubov theory in continuum. *Phys. Rev. C* **2012**, *85*, 024312. [CrossRef]
31. Li, L.; Meng, J.; Ring, P.; Zhao, E.G.; Zhou, S.G. Odd Systems in Deformed Relativistic Hartree Bogoliubov Theory in Continuum. *Chin. Phys. Lett.* **2012**, *29*, 042101. [CrossRef]
32. Kucharek, H.; Ring, P.; Schuk, P.; Bengtsson, R.; Girod, M. Pairing properties of nuclear matter from the gogny force. *Phys. Lett. B* **1989**, *216*, 249. [CrossRef]
33. Geiger, H.; Nuttall, J.M. LVII. The ranges of the α particles from various radioactive substances and a relation between range and period of transformation. *Philos. Mag. Ser.* **1911**, *22*, 613. [CrossRef]
34. Deng, J.-G.; Zhang, H.-F.; Royer, G. Improved empirical formula for α-decay half-lives. *Phys. Rev. C* **2020**, *101*, 034307. [CrossRef]
35. Rashidpour, Z.; Naderi, D. An empirical formula for the alpha decay half-lives. *Int. J. Mod. Phys. E* **2023**, *32*, 2350028. [CrossRef]
36. Luo, S.; Xu, Y.-Y.; Zhu, D.-X.; He, B.; Chu, P.-C.; Li, X.-H. Improved Geiger–Nuttall law for α-decay half-lives of heavy and superheavy nuclei. *Eur. Phys. J. A* **2022**, *58*, 244. [CrossRef]
37. Wang, M.; Huang, W.J.; Kondev, F.G.; Audi, G.; Naimi, S. The AME 2020 atomic mass evaluation (II). Tables, graphs and references. *Chin. Phys. C* **2021**, *45*, 030003. [CrossRef]
38. NNDC (National Nuclear Data Center). Brookhaven National Laboratory. Available online: https://www.nndc.bnl.gov/nudat2/ (accessed on 2 April 2025).
39. Gurvitz, S.A.; Kalbermann, G. Decay width and the shift of a quasistationary state. *Phys. Rev. Lett.* **1987**, *59*, 262. [CrossRef]
40. Xu, C.; Ren, Z. New deformed model of α-decay half-lives with a microscopic potential. *Phys. Rev. C* **2006**, *73*, 041301. [CrossRef]

particles

MDPI

Article

Shell Structure Evolution of U, Pu, and Cm Isotopes with Deformed Relativistic Hartree–Bogoliubov Theory in a Continuum

Liang Wu [1], Wei Zhang [1,*], Jing Peng [2,*] and Jinke Huang [1,3]

1 School of Physics, Zhengzhou University, Zhengzhou 450001, China
2 Department of Physics, Beijing Normal University, Beijing 100875, China
3 Haimen Experimental School, Nantong 226199, China
* Correspondence: zw76@pku.org.cn (W.Z.); jpeng@bnu.edu.cn (J.P.)

Abstract: By adopting the deformed relativistic Hartree–Bogoliubov theory in continuum (DRHBc) with the point-coupling density functional PC-PK1, we investigate the shell structure evolution of even–even U, Pu, and Cm isotopic chains from the proton drip line to the neutron drip line. The Fermi energy λ_n, two-neutron separation energy S_{2n}, two-neutron shell gap δ_{2n}, and quadrupole deformation β_2 all indicate the major shell closures at N = 126, 184, and 258. The emergence of sudden drops between U and Pu isotopic chains in the proton Fermi energies λ_p around these neutron shell closures is a consequence of the designation convention when the pairing collapse at the spurious shell closure Z = 92 occurs. The fine structure in the two-neutron shell gap, like negative δ_{2n}, may be related to the ground-state shape transition. Finally, the subshells indicated by the small-scale peaks in the two-neutron shell gaps can be well understood by the deformed gaps in the single-neutron levels obtained by DRHBc theory.

Keywords: DRHBc; shell structure; neutron drip line; single-particle levels; deformed shell

1. Introduction

The shell structure evolution of nuclei has been extensively studied in many efforts, playing an important role in nuclear physics and astrophysics [1,2]. Currently, many nuclear experimental facilities aiming at the exploration of exotic nuclei away from the β-stability valley are in use or under construction, such as the High-Intensity Heavy Ion Accelerator Facility (HIAF) in China [3], the Radioactive Ion Beam Factory (RIBF) in Japan [4], and the Facility for Rare Isotope Beams (FRIB) in the United States [5]. Over the years, experiments have led to a series of discoveries concerning the occurrence of new magic numbers and the disappearance of traditional magic numbers, for instance, evidence for shell closures at $N = 32, 34$ in ^{54}Ti [6] and ^{54}Ga [7], and the local $N = 40$ shell closure in the magic Ni isotopic chain [8].

Theoretically, the density functional theory provides a microscopic and self-consistent way to globally study the structure of atomic nuclei across the nuclear landscape [9,10]. For non-relativistic approaches, the calculation mass table has been conducted using Skyrme [11–14] or Gogny [15–17] Hartree–Fock–Bogoliubov density functional theories. For relativistic approaches, covariant density functional theory (CDFT) can provide a self-consistent description and has made significant advances as well [10,18–24].

Based on CDFT, considering pairing correlations, continuum effects, and axial deformation degrees of freedom simultaneously, the deformed relativistic Hartree–Bogoliubov theory in continuum (DRHBc) was developed [25,26]. Currently, DRHBc theory, grounded on the point-coupling density functional PC-PK1 [27], has evolved strategies and technologies for even-Z nuclei [28,29]. It has recently been applied to construct mass tables for even-Z nuclei in Refs. [30,31], and mass tables for odd-Z nuclei are currently in development. Many interesting topics have been discussed along this line. Specifically, DRHBc theory has been used to study the ground state properties of $Z = 6$ isotopic chains, with a particular focus on the study of the neutron halo phenomenon [32]. In addition, DRHBc theory has been applied to the research of the neutron drip line nuclei of superheavy nuclear regions with $Z = 106, 108, 110$; even–even neutron-rich regions with $Z = 8$ to $Z = 20$; and the position of neutron drip line nuclei for isotopic chains with $Z = 50$ to $Z = 70$ [33–35]. Moreover, DRHBc theory has successfully described the odd–even staggering and the kink structure in the charge radii of mercury isotopic chains [36]. It is worth noting that DRHBc theory has also been used to explore changes in quadrupole deformation with neutron number in Te, Xe, and Ba isotopic chains, and an interesting phenomenon is that as the neutron increases, all these isotopic chains exhibit a common characteristic of prolate-shape dominance [37]. Additionally, DRHBc theory has been used for a systematic study of shell structures. It not only successfully reproduces experimental data [38] and verifies the magic numbers but also reproduces the disappearance of traditional magic numbers and the emergence of new magic numbers [39]. The inner fission barriers of even–even uranium isotopes are studied studies using DRHBc theory, and a periodic-like behavior is exhibited: peaks at the shell closures and valleys in the mid-shells [40]. These research achievements fully demonstrate the wide applicability and accuracy of DRHBc theory in nuclear physics and astrophysics research.

In this paper, we will focus on the shell structure evolution in the even–even U, Pu, and Cm isotopic chains by adopting DRHBc theory. It is interesting to note that in a very recent experimental work [41], the proton drip line of their odd-Z neighboring element Np has been determined to be ^{219}Np, which marks the heaviest proton drip line ever reached. The established α-decay systematics suggested the robustness of the $N = 126$ shell closure even in the vicinity of the proton drip line [41]. Besides, it is fascinating to explore the possible shell closures, subshell structures, and their evolution in the neutron-rich region.

2. Theoretical Framework

The details of DRHBc theory can be found in Refs. [26,28,30,42]. Here, we only briefly introduce its formalism. In DRHBc theory, the relativistic Hartree–Bogoliubov (RHB) equation that self-consistently describes mean field potentials and pairing correlations reads

$$\begin{pmatrix} \hat{h}_D - \lambda_\tau & \hat{\Delta} \\ -\hat{\Delta}^* & -\hat{h}_D^* + \lambda_\tau \end{pmatrix} \begin{pmatrix} U_k \\ V_k \end{pmatrix} = E_k \begin{pmatrix} U_k \\ V_k \end{pmatrix}, \tag{1}$$

where $\hat{h}_D$ is the Dirac Hamiltonian, λ_τ $(\tau = n, p)$ is the Fermi energy for neutron or proton, $\hat{\Delta}$ is the pairing potential, U_k and V_k are the quasiparticle wave functions, and E_k is the quasiparticle energy. In the coordinate space, the Dirac Hamiltonian is

$$h_D(\boldsymbol{r}) = \boldsymbol{\alpha} \cdot \boldsymbol{p} + V(\boldsymbol{r}) + \beta[M + S(\boldsymbol{r})], \tag{2}$$

where M is the nucleon mass, $S(\boldsymbol{r})$ and $V(\boldsymbol{r})$ are the scalar and vector potentials. The pairing potential is

$$\Delta(\boldsymbol{r}_1, \boldsymbol{r}_2) = V^{pp}(\boldsymbol{r}_1, \boldsymbol{r}_2)\kappa(\boldsymbol{r}_1, \boldsymbol{r}_2), \tag{3}$$

where κ is the pairing tensor and V^{pp} is the pairing force of a density-dependent zero-range type,

$$V^{pp}(\boldsymbol{r}_1, \boldsymbol{r}_2) = V_0 \frac{1}{2}(1 - P^\sigma)\delta(\boldsymbol{r}_1 - \boldsymbol{r}_2)\left(1 - \frac{\rho(\boldsymbol{r}_1)}{\rho_{\text{sat}}}\right), \tag{4}$$

with V_0 as the pairing strength, $(1 - P^{\sigma})/2$ as the projector for the spin $S = 0$ component in the pairing channel, and ρ_{sat} as the saturation density of nuclear matter.

For an axially deformed nucleus with spatial reflection symmetry, the potentials and densities can be expanded in terms of Legendre polynomials,

$$f(r) = \sum_{\lambda} f_{\lambda}(r) P_{\lambda}(\cos\theta), \quad \lambda = 0, 2, 4, \ldots, \lambda_{\max} \tag{5}$$

Finally, in DRHBc theory, the RHB equations are solved using the basis expansion method with the Dirac WS basis [25,43,44], which can properly describe the large spatial extension of weakly bound nuclei.

3. Results and Discussion

Upon the efforts of the DRHBc mass table collaboration, the mass table for even-Z nuclei in DRHBc with the PC-PK1 density functional [27] has been constructed [29–31]. The numerical details adopted here for the U, Pu, and Cm isotopic chains follow the DRHBc mass table calculations for the nuclear region $72 \leqslant Z \leqslant 100$ [28,30]. That is, the box size $R_{\text{box}} = 20$ fm and the energy cutoff $E_{\text{cut}} = 300$ MeV for the Dirac WS basis have been chosen. The pairing strength $V_0 = -325\ \text{MeV} \cdot \text{fm}^3$ and the pairing window 100 MeV are used in DRHBc theory calculations. The Legendre expansion truncation order $\lambda_{\max} = 8$ and the angular momentum cutoff $J_{\max} = 23/2\hbar$ are taken.

First of all, for these three isotopic chains, the proton and neutron drip line nuclei will be determined by the Fermi energies and the two nucleon separation energies. The Fermi energy represents the change in total energy with the change in the number of particles [45]. A negative Fermi energy usually corresponds to a positive separation energy of a bound nucleus, and a positive Fermi energy implies an unbound nucleus in the mean field level [28]. The nucleon separation energy can provide direct information on whether the nucleus is able to emit nucleons or not.

Figure 1 shows the calculated neutron Fermi energies λ_{n} and proton Fermi energy λ_{p} for the even–even U, Pu, and Cm isotopic chains in DRHBc theory calculations. For the convention where the pairing energy is zero, the Fermi energy is chosen to be the energy of the last occupied single-particle state, which is the same strategy adopted in Refs. [28,29]. In Figure 1a, it is seen that the λ_{n} almost continuously increases with the neutron number for a whole isotopic chain and finally becomes positive when going beyond the neutron drip-line. The neutron drip-line nuclei for these isotopic chains are ^{350}U, ^{352}Pu, and ^{354}Cm, respectively, all corresponding to the neutron number $N = 258$. In Figure 1b, it is seen that the λ_{p} starts from a positive value and almost continuously decreases with the proton number for a whole isotopic chain. The first nucleus with negative λ_{p} corresponds to the proton drip line, which is ^{212}U with $N = 120$ ($\lambda_{\text{p}} = -0.16$ MeV) for U isotopes, ^{220}Pu with $N = 126$ ($\lambda_{\text{p}} = -0.07$ MeV) for Pu isotopes, and ^{226}Cm with $N = 130$ ($\lambda_{\text{p}} = -0.36$ MeV) for Cm isotopes. Corresponding, the three nuclei are the proton drip-line nuclei. Note that these neutron and proton drip-line nuclei determined by the Fermi energies are consistent with the results by the two-neutron separation energies S_{2n}.

The sudden changes in the Fermi energies may suggest the possible shell closures. In Figure 1a, the λ_{n} increases rapidly at certain neutron numbers, such as the well-known magic number $N = 126$ in U and Pu isotopic chains and $N = 184$ and 258 in U, Pu, and Cm isotopic chains. The latter indicates new shell closures at $N = 184$ and 258. In Figure 1b, in contrast to the Pu and Cm isotopic chains, the proton Fermi energy λ_{p} in the U isotopic chain shows sudden drops when approaching the possible neutron shell closures $N = 126$, 184 and 258. As interpreted below, these sudden drops can be attributed to the proton pairing collapse originating from the spurious shell closure at $Z = 92$ in the single-proton energy levels.

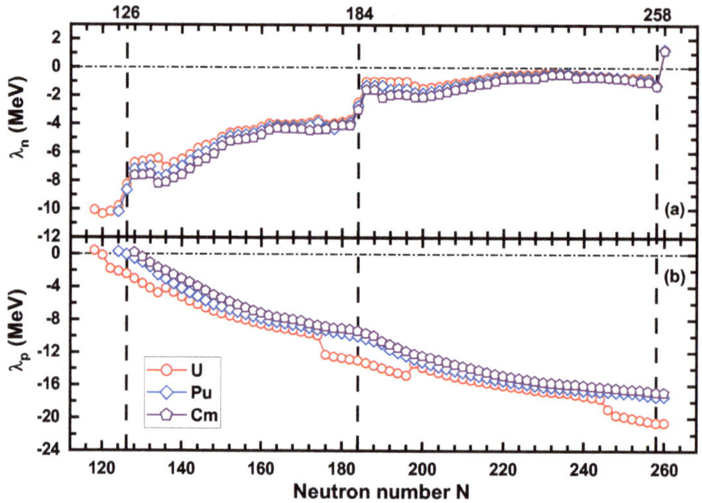

Figure 1. Neutron (**a**) and proton (**b**) Fermi energies of even−even U, Pu, and Cm isotopic chains in DRHBc theory calculations.

Taking the nuclei ^{276}U, ^{278}Pu, ^{280}Cm with $N = 184$ as an example, Figure 2 shows the single-proton levels near the Fermi energy as a function of the deformation parameter β_2. In this subfigure (a), there is a significant spherical shell gap of $Z = 92$ around 3.816 MeV between the degenerate spherical orbitals $2f_{7/2}$ and $1h_{9/2}$, which is even larger than the traditional spherical shell of $Z = 82$, as shown below. Due to the existence of this proton shell $Z = 92$, the U isotopes with neutron number not far away from shell closures tend to be spherical, and the corresponding proton pairing energy tends to be zero. This pairing collapse leads the proton Fermi energies around $N = 184$ for these Uranium isotopes located at the designated spherical orbital $1h_{9/2}$. For the neighboring uranium isotopes away from $N = 184$, the ground states are deformed with Fermi energies above $1h_{9/2}$. As a result, in Figure 1b, focusing the neutron numbers around the possible shell closure $N = 184$, there is one sudden drop from 174 to 176, jumping from Fermi energies above $1h_{9/2}$ in deformed nuclei to exactly $1h_{9/2}$ when pairing collapse occurs, together with one sudden rise from 196 to 198 when shape transition from spherical to deformed occurs. The shape evolution will be further discussed with quadrupole deformation later. For Pu and Cm isotopes with two or four more protons, in Figure 2b,c, the Fermi energies just stay between these two orbitals $1h_{9/2}$ and $2f_{7/2}$, both for spherical ground states close to $N = 184$ and for deformed ground states away from 184. Consequently, the Fermi energies around $N = 184$ for Pu and Cm isotopes displayed continuous behavior without abrupt changes as shown in Figure 1b. This spurious shell closure $Z = 92$, which is commonly found in the CDFT calculations, is expected to be cured or mitigated by the presence of tensor interaction [46] or localized Fock term [47].

From the binding energies, the two-neutron separation energies S_{2n} can be calculated as

$$S_{2n}(Z, N) = E_B(Z, N) - E_B(Z, N-2). \tag{6}$$

In Figure 3a, the S_{2n} values in the U, Pu, and Cm isotopic chains obtained from the DRHBc calculations are shown, in comparison with the RCHB results [48] and the available experimental data [38]. Around the shell closure $N = 126$, the results of the DRHBc and the

RCHB are consistent and reproduce its robustness in the vicinity of proton drip line [41]. Away from the shell-closure region, compared to the RCHB results, the DRHBc calculations reproduce the experimental values better. When the S_{2n} reaches a negative value, the nucleus is unbound against two-neutron emission. The last nucleus with the positive S_{2n} is determined as the two-neutron drip line nucleus. In the even–even U, Pu, and Cm isotopic chains, the last positive values are 1.41 MeV of ^{350}U, 1.73 MeV of ^{352}Pu, and 2.00 MeV of ^{354}Cm; thus, these three nuclei are at the two-neutron drip line $N = 258$. This is consistent with the position of the neutron drip line determined by the λ_n in Figure 1a.

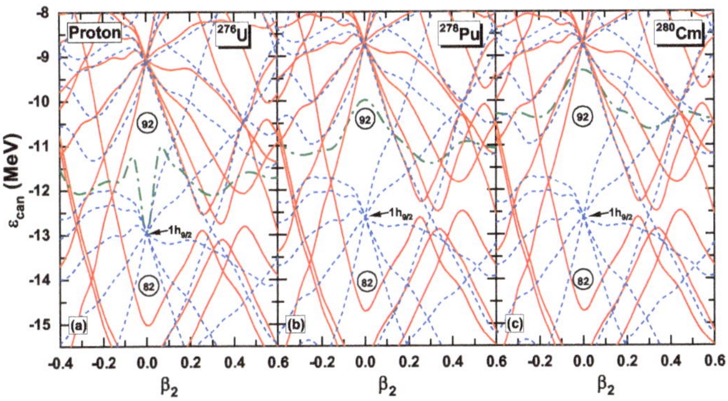

Figure 2. Single−proton energies in the canonical basis of ^{276}U (**a**), ^{278}Pu (**b**) , ^{280}Cm (**c**) as a function of the deformation parameter β_2 in DRHBc theory calculations. The Fermi energy as a function of quadrupole deformation is also displayed by a green dash−dot line. The single-particle levels with positive (negative) parity are displayed as red solid (blue dashed) lines. The proton number obtained by filling all the lower levels is shown with a circle at several energy gaps.

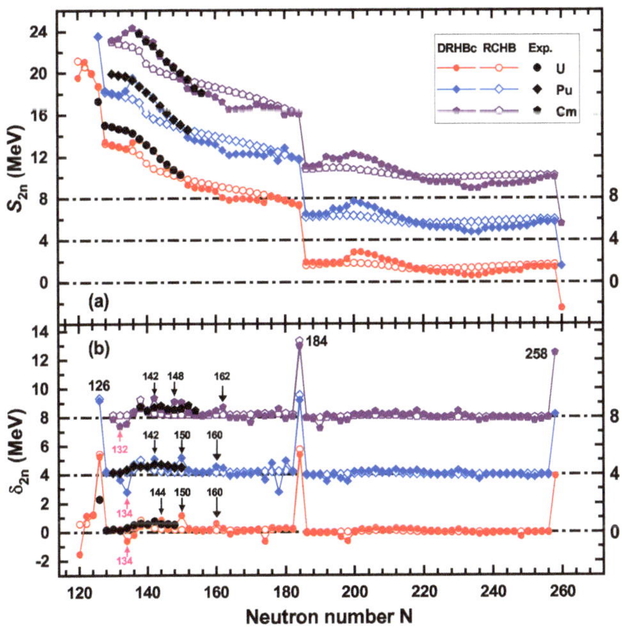

Figure 3. Two−neutron separation energies (**a**) and two−neutron shell gaps (**b**) of even−even U, Pu (shifted up 4 MeV), and Cm (shifted up 8 MeV) isotopic chains in DRHBc theory calculations. The RCHB results [48] and the available experimental data [38] are shown for comparison.

In Figure 3a, between the traditional shell closure $N = 126$ and two possible ones $N = 184$ and 258, comparing with the RCHB results, the S_{2n} from DRHBc theory exhibits systematic behavior. When the neutron number increases from the shell closure to the middle shell, S_{2n} from DRHBc theory is higher than that of RCHB calculation, while from the mid-shell to the next shell closure, S_{2n} from DRHBc theory is lower than that of RCHB calculation. It is noteworthy that the RCHB theory is spherical, while DRHBc theory takes the quadrupole deformation into account. Such behavior should be associated with the evolution of the ground state deformation.

It is further noted that some bound nuclei beyond the neutron drip line have been predicted in the DRHBc mass table calculations [30,31], such as in the $50 \leqslant Z \leqslant 70$ [35] and $106 \leqslant Z \leqslant 112$ [33,49,50] regions, which form the peninsulas of stability in the nuclear landscape. However, for the current region, after examining the neutron Fermi energies and the two-neutron separation energies for nuclei beyond the neutron drip line, this phenomenon of entrant stability is not found.

Besides the information of drip line, the S_{2n} contains detailed information about the shell closure. For example, as shown in Figure 3a, the S_{2n} in the U, Pu, and Cm isotopic chains show sudden drops at the traditional magic number $N = 126$ and possible magic numbers $N = 184$ and 258. Away from the shell closures, as the number of neutrons increases, the S_{2n} drops almost smoothly. However, near $N = 134$ and $N = 198$, the S_{2n} calculated using DRHBc theory shows a slight increase with the neutron number, whereas the S_{2n} obtained from RCHB theory does not exhibit the increasing behavior. Similar increases of S_{2n} are also observed in the DRHBc results at positions such as near $N = 174$, 248 in the U isotopic chain; $N = 174$, 178, 192, 236 in the Pu isotopic chain; and $N = 170$, 190 in the Cm isotopic chain. It is noteworthy that RCHB theory is spherical, while DRHBc theory takes the quadrupole deformation into account. Thus the quadrupole deformation may lead to the occurrence of these phenomena.

In order to more clearly demonstrate the shell closures and the fine structure of S_{2n} in the U, Pu, and Cm isotopic chains, the two-neutron shell gaps δ_{2n}, which are the difference of S_{2n} between two neighboring nuclei

$$\delta_{2n}(Z, N) = S_{2n}(Z, N) - S_{2n}(Z, N + 2), \tag{7}$$

are analyzed. Figure 3b shows the δ_{2n} of the even–even U, Pu, and Cm isotopic chains calculated by DRHBc theory with the neutron number together with the RCHB results [48] and the available experimental data [38]. Comparing with the RCHB results, DRHBc theory better reproduces the available experimental data. At the shell closures or possible shell closures $N = 126$, 184 and 258, the δ_{2n} obtained from both the DRHBc and RCHB calculations always exhibits pronounced peaks in the U, Pu, and Cm isotopic chains. As shown in Figure 3b, most δ_{2n} values, whether from the RCHB calculations or the DRHBc calculations, are non-negative, reflecting the globally declining behavior of S_{2n} with the increase in neutron number. The few negative δ_{2n} values appear only in the DRHBc results, the positions of which are around $N = 134$ and 198 for the three isotopic chains and also at $N = 174$, 248 for U, $N = 174$, 178, 192, 236 for Pu, and $N = 170$, 190 for Cm. In comparison, the δ_{2n} is always greater than zero for the RCHB results, which does not consider the quadrupole deformation degree of freedom. Therefore, it is necessary to examine the quadrupole deformation evolutions in these isotopic chains.

Figure 4 presents the ground state deformation evolutions of the U, Pu, and Cm isotopic chains obtained from the DRHBc calculations, in comparison with the available empirical data extracted from the observed $B(\mathrm{E2}, 0_1^+ \to 2_1^+)$ [38]. DRHBc theory can reproduce the available data. As seen in Figure 4, the deformation evolutions of the even–even U, Pu, and Cm isotopic chains show similar evolutionary patterns: near the well-known shell closure at $N = 126$, the ground state shapes are spherical or nearly spherical. Before reaching the next shell closure at $N = 184$, β_2 reaches a peak in the mid-shell. As the neutron number increases to $N = 172$, ^{268}Cm is the first to transit from prolate to oblate shape, followed by ^{268}Pu at $N = 174$ and ^{268}U at $N = 176$. From $N = 184$ to possible shell closure at $N = 258$, they all first experience a transition from spherical to prolate, with β_2 reaching a peak around $N = 214$. The U, Pu, and Cm isotopic chains experience transitions from prolate to oblate shape at $N = 236$, $N = 238$, and $N = 238$, respectively. After that, the ground state shape becomes less oblate with increasing neutron number, and finally becomes spherical at $N = 258$.

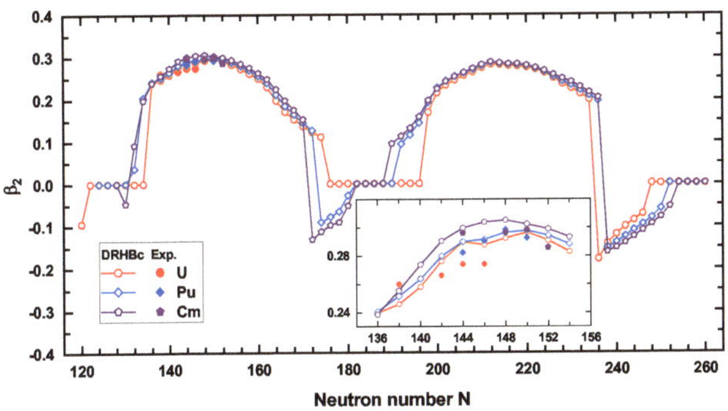

Figure 4. Quadrupole deformation of even−even U, Pu, and Cm isotopic chains in DRHBc theory calculations, and the available experimental data [38] are shown for comparison. The inset gives a detailed comparison between the calculated results and the available experimental data [38].

One may notice that the increase in S_{2n} in Figure 3a and the negative δ_{2n} in Figure 3b appear where the quadrupole deformation transition happens in Figure 4. For example, the negative δ_{2n} occurs at N = 134 for U and Pu isotopes, and N = 132 for Cm isotopes; correspondingly the shape transition from spherical to prolate occurs synchronically in Figure 4.

Besides the pronounced peaks of δ_{2n} that show the major shells at $N = 126$, 184, and 258, in Figure 3b, one can also find some small-scale peaks in the DRHBc results. The amplitudes of these peaks are about 1 MeV or less, and their positions are not strictly aligned for different isotopic chains. For uranium, these small peaks appear at N = 144, 150, and 160; for plutonium, at N = 142, 150, and 160; and for curium, at N = 142, 148, and 162. Note that these peaks do not appear in the RCHB results, but instead another small peak at $N = 138$ appears.

To understand the origins of these small peaks in Figure 3b, by comparing the U, Pu, and Cm isotones, Figure 5 shows the canonical single-neutron energies of ^{234,242,252}U, ^{236,244,254}Pu and ^{238,246,256}Cm as a function of the deformation parameter β_2 obtained by DRHBc theory. When the nucleus is spherical at $\beta_2 = 0$, the neutron single particle levels with the same quantum numbers nlj are degenerate due to the spherical symmetry, leading to several energy gaps between two neighboring spherical orbitals. The energy gap between the spherical orbitals $1i_{11/2}$ and $2g_{9/2}$ is around 2 MeV. With the $1i_{11/2}$ orbital fully occupied by 12 neutrons lying above the shell closure N = 126, a spherical gap at N = 138 is formed. This can be used to interpret the small peak of δ_{2n} at N = 138 shown by the spherical RCHB results in Figure 3.

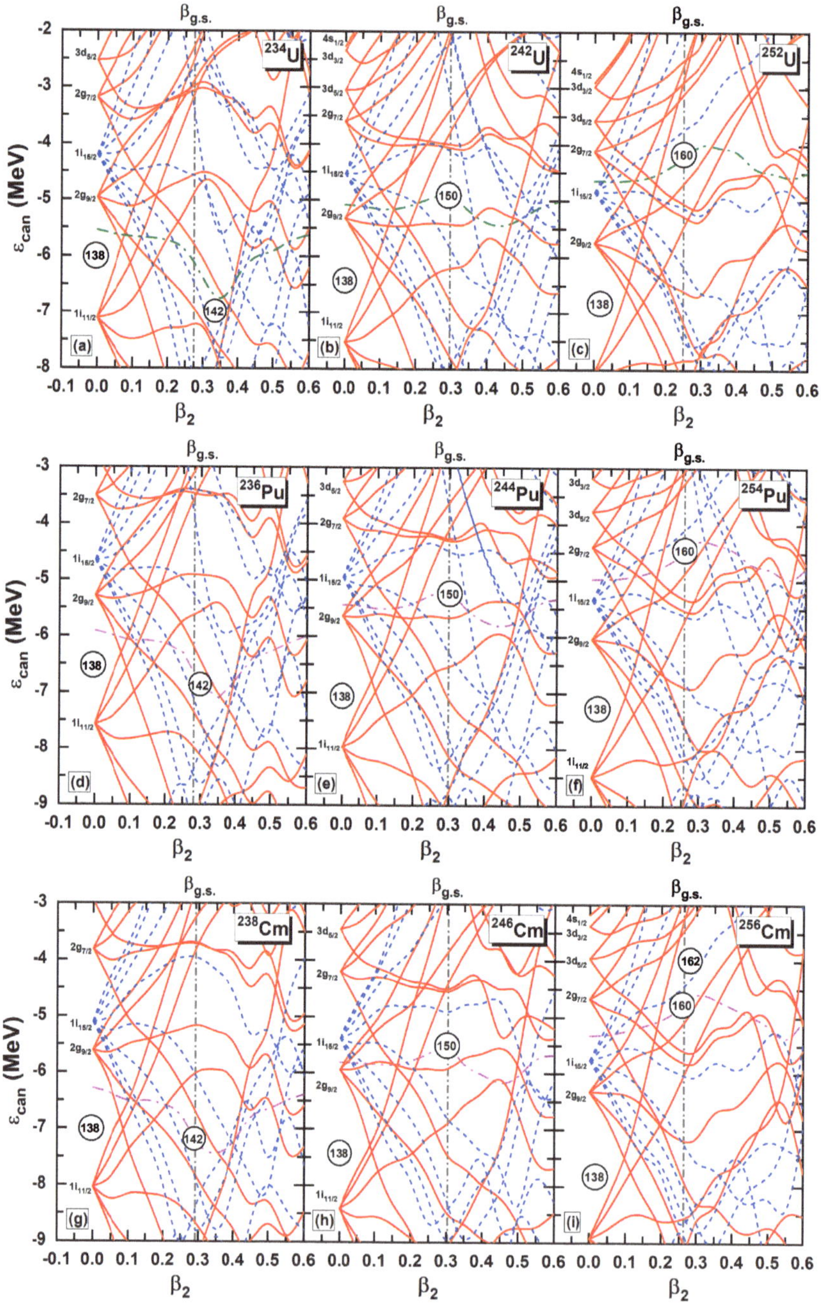

Figure 5. Single−neutron energies in the canonical basis for uranium, plutonium and curium isotopes with N = 142, 150 and 160, namely ^{234,242,252}U (**a**–**c**), ^{236,244,254}Pu (**d**–**f**), and ^{238,246,256}Cm (**g**–**i**), as functions of the deformation parameter β_2 in DRHBc theory. The Fermi energy is also displayed by a green dash−dot line. The single−particle levels with positive (negative) parity are displayed as red solid (blue dashed) lines. The neutron numbers obtained by filling all the lower levels are shown in a circle at several energy gaps. The ground states are indicated by gray lines at corresponding deformation positions.

When the spherical symmetry is broken and a nucleus undergoes axial deformation, the total angular momentum j is no longer a good quantum number while its projection on the symmetry axis K remains a good quantum number. Normally, one spherical j-shell splits into $(2j+1)/2$ orbitals. As a result, the traditional spherical shell closure breaks down and new deformed subshells may emerge. As shown in Figure 5, comparing panels (a), (d), and (g), a shell gap at 142 appears obviously for the $N=142$ isotones ^{236}Pu and ^{238}Cm but not so distinctly for $N=142$ isotone ^{234}U. Instead, for ^{234}U, a gap slightly larger appears at $N=144$, just above 142. This gap change effectively explains the subtle peak shift from $N=144$ for U to $N=142$ for Cm in Figure 3b. Similarly, a gap at 160 emerges for $N=160$ isotones ^{252}U and ^{254}Pu in panel (c) and (f), while a gap at 162 is seen for ^{256}Cm in panel (i), corresponding to the peak shift from $N=160$ for U and Pu to $N=162$ for Cm in Figure 3b. For panels (b) ,(e), and (h), a relatively robust subshell gap at $N=150$ can be found for ^{242}U, ^{244}Pu, and ^{246}Cm, which generally agrees with the peak at $N=150$ for U, Pu in Figure 3b. Based on the analysis, the small peak of two neutron gap δ_{2n} strongly correlates with the deformed subshell structure near the Fermi surface, and some subshells may be quenched with different isotopic chains.

Another important indicator of shell closure is the average pairing gap, as discussed in Ref. [51], which allows us to obtain direct information about the impact of pairing correlations and shows the arch structures that vanish at shell closures and have additional deeps at subshell closures. Figure 6a shows the Δ_n as a function of the number of neutrons for even–even U, Pu, and Cm isotopic chains calculated by DRHBc theory. As shown in Figure 6a, most Δ_n values transition smoothly with the neutron number and are greater than zero. However, near the magic number N = 126 and the possible magic numbers N = 184 and 258, Δ_n values exhibit a sharp decrease and become zero. Additionally, we observe that a few Δ_n values also show a decrease at the following positions: for the U isotopic chain, N = 144, 150, and 160; for the Pu isotopic chain, N = 142, 150, and 160; and for the Cm isotopic chain, N = 142, 148, and 162. These positions correspond to the small-scale peaks in Figure 3b.

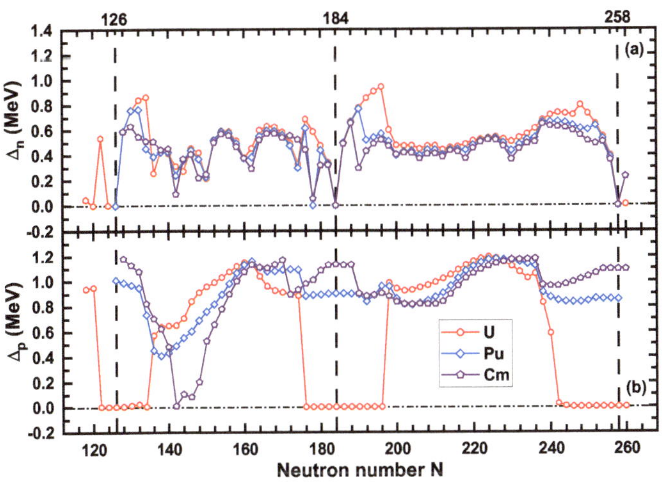

Figure 6. Neutron average pairing gap Δ_n (**a**) and proton average pairing gap Δ_p (**b**) of even−even U, Pu, and Cm isotopic chains in DRHBc theory calculations.

Figure 6b shows the Δ_p as a function of the number of protons for even–even U, Pu, and Cm isotopic chains calculated by DRHBc theory. Similarly, in Figure 6b, most Δ_p values exceed zero. It is observed that for the U isotopic chain, Δ_p values drop to zero near the magic number $N = 126$ and the possible magic numbers $N = 184$ and 258. In contrast, the Δ_p values for Pu and Cm isotopic chains remain positive. This indicates that due to the existence of the pseudo shell $Z = 92$, a pairing collapse occurs in the U isotopic chain within these regions, resulting in a sudden decrease in the λ_p value in the U isotopic chain shown in Figure 1.

Very recently, Ref. [52] proposed a new binding-energy indicator, single-particle neutron energy Δe_n , to catalog the subshell closures in nuclear landscape. This indicator behaves in a similar way as δ_{2n}. The subshell closures at $N = 152$ and 162 are found for different mass models SkM*, SLy4, UNEDF1, UNEDF2, and FRDM-2012. These two subshells are very close to our results presented here.

4. Conclusions

In summary, we have performed systematic studies of DRHBc theory with the PC-PK1 density functional for the shell structure evolution of even–even U, Pu, and Cm isotopic chains from the proton drip line to the neutron drip line. By analyzing the Fermi energies and the two-neutron separation energies, we predicted the proton drip line for the U, Pu, and Cm isotopic chains to be at ^{212}U, ^{220}Pu, and ^{226}Cm and the neutron drip line to be at ^{350}U, ^{352}Pu, and ^{354}Cm, respectively. In comparison with spherical RCHB calculations, the inclusion of axial deformation degree of freedom does not change these drip line boundaries. The phenomenon of entrant stability beyond the neutron drip line is not found for the current region. The shell closures at $N = 126$, 184, and 258 in U, Pu, and Cm isotopic chains can be clearly seen by the Fermi energy λ_n, two-neutron separation energy S_{2n}, and two-neutron shell gap δ_{2n}. The emergence of sudden drops between U and Pu isotopic chains in the proton Fermi energies λ_p around some neutron shell closures $N = 126$, 184, and 258 is a consequence of designation convention when the pairing collapse at the spurious shell closure $Z = 92$ occurs.

The fine structure in the two-neutron separation energy and the two-neutron shell gap, like decrease of S_{2n} and negative δ_{2n}, may be related to the ground state shape transition.

Finally, the small-scale peaks in the two-neutron shell gaps δ_{2n} indicate some subshells, namely, $N = 144$, 150, and 160 for uranium, $N = 142$, 150, and 160 for plutonium, and $N = 142$, 148, and 162 for curium. These subshells can be well understood by the deformed gaps in the single-neutron levels obtained by DRHBc theory.

Author Contributions: Conceptualization, W.Z. and J.P.; methodology, L.W.; software, J.H.; formal analysis, L.W.; investigation, L.W.; resources, W.Z.; data curation, J.H.; writing—original draft preparation, L.W.; writing—review and editing, J.P.; visualization, L.W.; supervision, W.Z. All authors have read and agreed to the published version of the manuscript.

Funding: This work was partly supported by the Natural Science Foundation of Henan Province (Grant No. 242300421156, 202300410480), the National Natural Science Foundation of China (Grant No. U2032141), and the Super Computing Center of Beijing Normal University.

Data Availability Statement: The study does not report any finalized data yet.

Acknowledgments: All authors deeply thank Shuang-quan Zhangfor detailed discussion and revision. Helpful discussions with members of the DRHBc Mass Table Collaboration are highly appreciated.

Conflicts of Interest: The authors declare no conflicts of interest.

References

1. Sorlin, O.; Porquet, M.-G. Nuclear magic numbers: New features far from stability. *Prog. Part. Nucl. Phys.* **2008**, *61*, 602.
2. Otsuka, T.; Gade, A.; Sorlin, O.; Suzuki, T.; Utsuno, Y. Evolution of shell structure in exotic nuclei. *Rev. Mod. Phys.* **2020**, *92*, 015002.

3. Zhan, W.L.; Xu, H.S.; Xia, J.W.; Zhao, H.W.; Yuan, Y.J. Progress in HIRFL-CSR. *Nucl. Phys. A* **2010**, *834*, 694c.
4. Motobayashi, T. RIKEN RI Beam Factory-Recent Results and Perspectives. *Nucl. Phys. A* **2010**, *834*, 707c.
5. Thoennessen, M. Plans for the Facility for Rare Isotope Beams. *Nucl. Phys. A* **2010**, *834*, 688c.
6. Dinca, D.C.; Janssens, R.V.F.; Gade, A.; Bazin, D.; Broda, R.; Brown, B.A.; Campbell, C.M.; Carpenter, M.P.; Chowdhury, P.; Cook, J.M.; et al. Reduced transition probabilities to the first 2+ state in Ti-52,54,56 and development of shell closures at N=32,34. *Phys. Rev. C* **2005**, *71*, 041302.
7. Steppenbeck, D.; Takeuchi, S.; Aoi, N.; Doornenbal, P.; Matsushita, M.; Wang, H.; Baba, H.; Fukuda, N.; Go, S.; Honma, M.; et al. Evidence for a new nuclear 'magic number' from the level structure of ^{54}Ca. *Nature* **2013**, *502*, 207.
8. de Groote, R.P.; Billowes, J.; Binnersley, C.L.; Bissell, M.L.; Cocolios, T.E.; Day Goodacre, T.; Farooq-Smith, G.J.; Fedorov, D.V.; Flanagan, K.T.; Franchoo, S.; et al. Measurement and microscopic description of odd–even staggering of charge radii of exotic copper isotopes. *Nat. Phys.* **2020**, *16*, 620.
9. Bender, M.; Heenen P.H.. Self-consistent mean-field models for nuclear structure. *Rev. Mod. Phys.* **2003**, *75*, 121.
10. Meng, J.; Toki, H.; Zhou, S.G.; Zhang, S.Q.; Long, W.H.; Geng, L.S. Relativistic Continuum Hartree Bogoliubov theory for ground state properties of exotic nuclei. *Prog. Part. Nucl. Phys.* **2006**, *57*, 470.
11. Samyn, M.; Goriely, S.; Heenen, P.-H.; Pearson, J.M.; Tondeur, F. A Hartree–Fock–Bogoliubov mass formula. *Nucl. Phys. A* **2002**, *700*, 142.
12. Stoitsov, M.V.; Dobaczewski, J.; Nazarewicz, W.; Pittel, S.; Dean, D.J. Systematic study of deformed nuclei at the drip lines and beyond. *Phys. Rev. C* **2003**, *68*, 054312.
13. Goriely, S.; Chamel, N.; Pearson, J.M. Skyrme-Hartree–Fock–Bogoliubov nuclear mass formulas: Crossing the 0.6 MeV threshold with microscopically deduced pairing. *Phys. Rev. Lett.* **2009**, *102*, 152503.
14. Goriely, S.; Chamel, N.; Pearson, J.M. Further explorations of Skyrme-Hartree–Fock–Bogoliubov mass formulas. 13. The 2012 atomic mass evaluation and the symmetry coefficient. *Phys. Rev. C* **2013**, *88*, 024308.
15. Hilaire, S.; Girod, M. Large-scale mean-field calculations from proton to neutron drip lines using the D1S Gogny force. *Eur. Phys. J. A* **2007**, *33*, 237.
16. Goriely, S.; Hilaire, S.; Girod, M. First Gogny-Hartree–Fock–Bogoliubov Nuclear Mass Model. *Phys. Rev. Lett.* **2009**, *102*, 242501.
17. Delaroche, J.P.; Girod, M.; Libert, J.; Goutte, H.; Hilaire, S.; Peru, S.; Pillet, N.; Bertsch, G.F. Structure of even–even nuclei using a mapped collective Hamiltonian and the D1S Gogny interaction. *Phys. Rev. C* **2010**, *81*, 014303.
18. Ring, P. Relativistic mean field in finite nuclei. *Prog. Part. Nucl. Phys.* **1996**, *37*, 193.
19. Vretenar, D.; Afanasjev, A.V.; Lalazissis, G.A.; Ring, P. Relativistic Hartree Bogoliubov theory: static and dynamic aspects of exotic nuclear structure. *Phys. Rep.* **2005**, *409*, 101.
20. Niksic, T.; Vretenar, D.; Ring, P. Relativistic Nuclear Energy Density Functionals: Mean-Field and Beyond. *Prog. Part. Nucl. Phys.* **2011**, *66*, 519.
21. Meng, J.; Peng, J.; Zhang, S.Q.; Zhao, P.W. Progress on tilted axis cranking covariant density functional theory for nuclear magnetic and antimagnetic rotation. *Front. Phys.* **2013**, *8*, 55.
22. Meng, J.; Zhou, S.G. Halos in medium-heavy and heavy nuclei with covariant density functional theory in continuum. *J. Phys. G* **2015**, *42*, 093101.
23. Zhou, S.G. Multidimensionally constrained covariant density functional theories-nuclear shapes and potential energy surfaces. *Phys. Scr.* **2016**, *91*, 063008.
24. Shen, S.H.; Liang, H.Z.; Long, W.H.; Meng, J.; Ring, P. Towards an *abinitio* covariant density functional theory for nuclear structure. *Prog. Part. Nucl. Phys.* **2019**, *109*, 103713.
25. Zhou, S.G.; Meng, J.; Ring, P.; Zhao, E.G. Neutron halo in deformed nuclei. *Phys. Rev. C* **2010**, *82*, 011301.
26. Li, L.L.; Meng, J.; Ring, P.; Zhao, E.G.; Zhou, S.G. Deformed relativistic Hartree Bogoliubov theory in continuum. *Phys. Rev. C* **2012**, *85*, 024312.
27. Zhao, P.W.; Li, Z.P.; Yao, J.M.; Meng, J. New parametrization for the nuclear covariant energy density functional with a point-coupling interaction. *Phys. Rev. C* **2010**, *82*, 054319.
28. Zhang, K.Y.; Cheoun, M.K.; Choi, Y.B.; Chong, P.S.; Dong, J.M.; Geng, L.S.; Ha, E.; He, X.T.; Heo, C.; Ho, M.C.; et al. Deformed relativistic Hartree–Bogoliubov theory in continuum with a point-coupling functional: Examples of even–even Nd isotopes. *Phys. Rev. C* **2020**, *102*, 024314.
29. Pan, C.; Cheoun, M.K.; Choi, Y.B.; Dong, J.M.; Du, X.K.; Fan, X.H.; Gao, W.; Geng, L.S.; Ha, E.; He, X.T.; et al. Deformed relativistic Hartree–Bogoliubov theory in continuum with a point-coupling functional. II. Examples of odd Nd isotopes. *Phys. Rev. C* **2022**, *106*, 014316.
30. Zhang, K.Y.; Cheoun, M.K.; Choi, Y.B.; Chong, P.S.; Dong, J.M.; Dong, Z.H.; Du, X.K.; Geng, L.S.; Ha, E.; He, X.T.; et al. Nuclear mass table in deformed relativistic Hartree–Bogoliubov theory in continuum, I: Even–even nuclei. *At. Data Nucl. Data Tables* **2022**, *144*, 101488.

31. Guo, P.; Cao, K.; Chen, K.; Chen, Z.; Cheoun, M. K.; Choi, Y.B.; Lam, P.C.; Deng, W.; Dong, J.; Du, X.; et al. Nuclear mass table in deformed relativistic Hartree–Bogoliubov theory in continuum, II: Even-Z nuclei. *At. Data Nucl. Data Tables* **2024**, *158*, 101661.
32. Sun, X.X.; Zhao, J.; Zhou, S.G. Study of ground state properties of carbon isotopes with deformed relativistic Hartree–Bogoliubov theory in continuum. *Nucl. Phys. A* **2020**, *1003*, 122011.
33. He, X.T.; Wang, C.; Zhang, K.Y.; Shen, C.W. Possible existence of bound nuclei beyond neutron drip lines driven by deformation. *Chin. Phys. C* **2021**, *45*, 101001.
34. In, E.J.; Kim, Y.; Papakonstantinou, P.; Hong, S.W. Neutron drip line in the deformed relativistic Hartree–Bogoliubov theory in continuum: Oxygen to Calcium. *Int. J. Mod. Phys. E* **2021**, *30*, 2150009.
35. Pan, C.; Zhang, K.Y.; Chong, P.S.; Heo, C.; Ho, M.C.; Lee, J.; Li, Z.P.; Sun, W.; Tam, C.K.; Wong, S.H.; et al. Possible bound nuclei beyond the two-neutron drip line in the $50 \leq Z \leq 70$ region. *Phys. Rev. C* **2021**, *104*, 024331.
36. Mun, M.H.; Kim, S.; Cheoun, M.K.; So, W.Y.; Choi, S.; Ha, E. odd–even shape staggering and kink structure of charge radii of Hg isotopes by the deformed relativistic Hartree–Bogoliubov theory in continuum. *Phys. Lett. B* **2023**, *847*, 138298.
37. Guo, P.; Pan, C.; Zhao, Y.C.; Du, X.K.; Zhang, S.Q. Prolate-shape dominance in atomic nuclei within the deformed relativistic Hartree–Bogoliubov theory in continuum. *Phys. Rev. C* **2023**, *108*, 014319.
38. Wang, M.; Huang, W.J.; Kondev, F.G.; Audi, G.; Naimi, S. The AME 2020 atomic mass evaluation (II). Tables, graphs and references. *Chin. Phys. C* **2021**, *45*, 030003.
39. Zheng, R.Y.; Sun, X.X.; Shen, G.F.; Geng, L.S. Evolution of N = 20, 28, 50 shell closures in the $20 \leq Z \leq 30$ region in deformed relativistic Hartree–Bogoliubov theory in continuum. *Chin. Phys. C* **2024**, *48*, 014107.
40. Zhang, W.; Huang, J.K.; Sun, T.T.; Peng, J.; Zhang, S.Q. Inner fission barriers of uranium isotopes in the deformed relativistic Hartree–Bogoliubov theory in continuum. *Chin. Phys. C* **2024**, *48*, 104105.
41. Zhang, Z.Y.; Gan, Z.G.; Yang, H.B.; Ma, L.; Huang, M.H.; Yang, C.L.; Zhang, M.M.; Tian, Y.L.; Wang, Y.S.; Sun, M.D.; et al. New Isotope Np220 : Probing the Robustness of the N=126 Shell Closure in Neptunium. *Phys. Rev. Lett.* **2019**, *122*, 192503.
42. Li, L.L.; Meng, J.; Ring, P.; Zhao, E.G.; Zhou, S.G. Odd systems in deformed relativistic Hartree Bogoliubov theory in continuum. *Chin. Phys. Lett.* **2012**, *29*, 042101.
43. Zhang, K.Y.; Pan, C.; Zhang, S.Q. Optimized Dirac Woods-Saxon basis for covariant density functional theory. *Phys. Rev. C* **2022**, *106*, 024302.
44. Zhou, S.G.; Meng, J.; Ring, P. Spherical relativistic Hartree theory in a Woods-Saxon basis. *Phys. Rev. C* **2003**, *68*, 034323.
45. Ring, P.; Schuck, P. *The Nuclear Many-Body Problem*; Springer: Berlin/Heidelberg, Germany, 1980.
46. Liang, H.Z.; Meng, J.; Zhou, S.G. Hidden pseudospin and spin symmetries and their origins inatomic nuclei. *Phys. Rep.* **2015**, *570*, 1.
47. Zhao, Q.; Ren, Z.X.; Zhao, P.W.; Meng, J. Covariant density functional theory with localized exchange terms. *Phys. Rev. C* **2022**, *106*, 034315.
48. Xia, X.W.; Lim, Y.; Zhao, P.W.; Liang, H.Z.; Qu, X.Q.; Chen, Y.; Liu, H.; Zhang, L.F.; Zhang, S.Q.; Kim, Y.; et al. The limits of the nuclear landscape explored by the relativistic continuum Hartree–Bogoliubov theory. *At. Data Nucl. Data Tables* **2018**, *121*, 1.
49. Zhang, K.Y.; He, X.T.; Meng, J.; Pan, C.; Shen, C.W.; Wang, C.; Zhang, S.Q. Predictive power for superheavy nuclear mass and possible stability beyond the neutron drip line in deformed relativistic Hartree–Bogoliubov theory in continuum. *Phys. Rev. C* **2021**, *104*, L021301.
50. He, X.T.; Wu, J.W.; Zhang, K.Y.; Shen, C.W. odd–even differences in the stability peninsula in the $106 \leq Z \leq 112$ region with the deformed relativistic Hartree–Bogoliubov theory in continuum. *Phys. Rev. C* **2024**, *110*, 014301.
51. Sil, T.; Patra, S.K.; Sharma, B.K.; Centelles, M.; Viñas, X. Superheavy nuclei in a relativistic effective Lagrangian model. *Phys. Rev. C* **2004**, *69*, 044315.
52. Buskirk, L.; Godbey, K.; Nazarewicz, W.; Satula, W. Nucleonic shells and nuclear masses. *Phys. Rev. C* **2024**, *109*, 044311.

particles

MDPI

Article

Determining the Ground State for Superheavy Nuclei from the Deformed Relativistic Hartree–Bogoliubov Theory in Continuum

Sibo Wang [1,*], Peng Guo [2] and Cong Pan [3]

1 Chongqing Key Laboratory for Strongly Coupled Physics, Department of Physics, Chongqing University, Chongqing 401331, China
2 State Key Laboratory of Nuclear Physics and Technology, School of Physics, Peking University, Beijing 100871, China; 2301110125@pku.edu.cn
3 Department of Physics, Anhui Normal University, Wuhu 241000, China; cpan@ahnu.edu.cn
* Correspondence: sbwang@cqu.edu.cn

Abstract: The deformed relativistic Hartree–Bogoliubov theory in continuum (DRHBc) has garnered significant attention for its ability to describe the properties of nuclei across the entire nuclear chart, from light to heavy nuclei, including both stable and exotic ones. As part of ongoing efforts to construct a mass table using the DRHBc theory, determining the ground states of nuclei is a crucial task in the systematic studies of deformed nuclei. In this work, a strategy for identifying the ground state in the superheavy nuclei region is proposed and evaluated, by taking $Z = 134$ and 135 isotopes as examples. First, we examine how the step size of the initial quadrupole deformation parameter, $\Delta\beta_2$, affects the pattern of the potential energy curves (PECs) and the determination of the ground state. Our findings indicate that $\Delta\beta_2 = 0.05$ producing smooth and well-defined PECs while maintaining an acceptable numerical cost. Next, we explore the convergence of PECs with respect to the angular momentum cutoff, $J_{\max}$. Based on the results, we recommend using $J_{\max} = 31/2\hbar$, especially for nuclei with competing oblate and prolate minima. Finally, we conclude that the accurate identification of the ground state can be achieved by performing unconstrained calculations around the minima of the PECs.

Keywords: superheavy nuclei; deformed relativistic Hartree Bogoliubov theory in continuum; potential energy curve; deformation

1. Introduction

The importance of nuclear mass cannot be overstated in the realm of nuclear physics [1,2]. It serves as a fundamental property that influences various phenomena, including nuclear stability, decay processes, and the structure of atomic nuclei. Accurate knowledge of nuclear masses is essential for understanding the underlying nucleon–nucleon interactions and for predicting the behavior of isotopes in astrophysical environments, such as nucleosynthesis in stars [3,4] and stellar neutrino emission [5].

The nuclear mass of superheavy nuclei [6] with $Z \geq 104$ is particularly interesting, in the sense that the exploration of charge and mass limits of atomic nuclei and the synthesis of long-lived or stable superheavy nuclei are at the frontier of modern nuclear physics [7–9]. Determining the ground states of superheavy nuclei is exceptionally challenging due to their short-lived nature and the complexities associated with their production. Nevertheless, advancements in experimental techniques, such as the use of gas-filled separators [10] and advanced detection systems [11,12], have enabled significant progress in identifying and characterizing these elusive nuclei. The insights gained from these studies are crucial for understanding the limits of nuclear stability and exploring the location of the island of stability [13,14]. Experimentally, the element Og with proton number $Z = 118$ is the highest

Z element observed so far [15]. Although we have witnessed the prosperous development of new generations of radioactive ion beam facilities, most neutron-rich nuclei far from the stability valley will remain beyond experimental access in the foreseeable future.

A reliable theoretical nuclear mass table is highly desired to further understand the nuclear landscape. Lots of efforts towards precise descriptions of nuclear masses have been made with various macroscopic–microscopic models [16–18], nonrelativistic density functional theories (DFTs) with Skyrme [19,20] and Gogny [21] interactions, and relativistic DFTs [22–31]. Among them, the deformed relativistic Hartree–Bogoliubov theory in continuum (DRHBc) [32,33] with the PC-PK1 [34] density functional has shown its remarkable ability on the satisfactory description of the ground-state properties with powerful explorations [35], due to the self-consistent consideration of the nuclear superfluidity, deformation, and continuum effects. In particular, the DRHBc Mass Table Collaboration [36] represents a concerted effort not only to calculate masses for stable and unstable nuclei but also to provide a more complete picture of the nuclear landscape.

Systematic numerical convergence checks from light to heavy nuclei for the DRHBc calculations have been justified in refs. [33,37]. Following the strategy and techniques presented in those articles, the nuclear mass table calculated by the DRHBc theory with PC-PK1 has been constructed for even–even nuclei [28] and even-Z nuclei with $8 \leq Z \leq 120$ [29]. Recently, this collaboration has extended its research scope to heavier nuclei.

In this work, we focus on the application of the DRHBc theory in the superheavy region with $Z > 120$, especially the determination of the ground state. In the literature, those nuclei with $Z > 126$ are also called hyperheavy nuclei [38–40]. This paper is organized as follows: Section 2 provides a brief overview of the theoretical framework, while Section 3 presents the numerical details. The results and discussions are presented in Section 4, and a summary is given in Section 5.

2. Theoretical Framework

For the sake of completeness, we lay out some key elements of the DRHBc theory with point-coupling density functionals. For details on the theoretical framework, we refer the reader to refs. [32,33,37,41].

The relativistic Hartree–Bogoliubov (RHB) equation describing the motion of nucleons in nuclei reads

$$\begin{pmatrix} \hat{h}_D - \lambda_\tau & \hat{\Delta} \\ -\hat{\Delta}^* & -\hat{h}_D^* + \lambda_\tau \end{pmatrix} \begin{pmatrix} U_k \\ V_k \end{pmatrix} = E_k \begin{pmatrix} U_k \\ V_k \end{pmatrix}, \tag{1}$$

where $\hat{h}_D$ is the Dirac Hamiltonian, $\hat{\Delta}$ is the pairing field, λ_τ is the Fermi energy for neutron or proton ($\tau = n, p$), U_k, V_k are the quasiparticle wave functions, and E_k is the quasiparticle energy.

In a nuclear system with time-reversal symmetry, the Dirac Hamiltonian in the coordinate space is written as

$$h_D(\boldsymbol{r}) = \boldsymbol{\alpha} \cdot \boldsymbol{p} + V(\boldsymbol{r}) + \beta[M + S(\boldsymbol{r})]. \tag{2}$$

Here, M is the nucleon mass, $\boldsymbol{\alpha}$ and β are Dirac matrices, and $S(\boldsymbol{r})$ and $V(\boldsymbol{r})$ are the scalar and vector potentials. In the point-coupling framework [34], the scalar potential $S(\boldsymbol{r})$ and vector potential $V(\boldsymbol{r})$ are defined as follows

$$S(\boldsymbol{r}) = \alpha_S \rho_S + \beta_S \rho_S^2 + \gamma_S \rho_S^3 + \delta_S \Delta \rho_S, \tag{3a}$$

$$V(\boldsymbol{r}) = \alpha_V \rho_V + \gamma_V \rho_V^3 + \delta_V \Delta \rho_V + eA^0 + \alpha_{TV} \tau_3 \rho_3 + \delta_{TV} \tau_3 \Delta \rho_3, \tag{3b}$$

where A^0 is the electric potential, $\tau_3 = \pm 1$ for neutron and proton. The coupling constants α's, β's, γ's, and δ's are adjustable parameters in the point-coupling Lagrangian and can be determined by fitting the binding energies and charge radii of a set of spherical nuclei as in ref. [34]. The local densities ρ_S, ρ_V, and ρ_3 are calculated with the quasiparticle wave functions

$$\rho_S(\boldsymbol{r}) = \sum_{k>0} V_k^\dagger(\boldsymbol{r})\gamma^0 V_k(\boldsymbol{r}), \tag{4a}$$

$$\rho_V(\boldsymbol{r}) = \sum_{k>0} V_k^\dagger(\boldsymbol{r}) V_k(\boldsymbol{r}), \tag{4b}$$

$$\rho_3(\boldsymbol{r}) = \sum_{k>0} V_k^\dagger(\boldsymbol{r})\tau_3 V_k(\boldsymbol{r}). \tag{4c}$$

Note that the no-sea approximation is adopted in Equation (4), i.e., the summations are performed only over the quasiparticle states in the Fermi sea.

The pairing field in the RHB Equation (1) is expressed as [32,33]

$$\Delta(\boldsymbol{r}_1, \boldsymbol{r}_2) = V^{pp}(\boldsymbol{r}_1, \boldsymbol{r}_2)\kappa(\boldsymbol{r}_1, \boldsymbol{r}_2), \tag{5}$$

where the spin and isospin indexes are not shown for simplicity. The quantity κ is the pairing tensor and V^{pp} is the density-dependent zero-range pairing force

$$V^{pp}(\boldsymbol{r}_1, \boldsymbol{r}_2) = V_0 \frac{1}{2}(1 - P^\sigma)\delta(\boldsymbol{r}_1 - \boldsymbol{r}_2)[1 - \rho(\boldsymbol{r}_1)/\rho_{\text{sat}}]. \tag{6}$$

In Equation (6), V_0 is the pairing strength, $\rho_{\text{sat}} = 0.152\ \text{fm}^{-3}$ is the saturation density of nuclear matter, $(1 - P^\sigma)/2$ is the projector for the spin-zero component in the pairing channel.

A deformed model is employed here because it enables us to determine the shape of nuclei, whether they are spherical or axially deformed, based on the total energies. This is definitely important because most studies on hyperheavy nuclei within DFTs [38,39,42,43] have been performed only for spherical shapes, while there is no guarantee that spherical minimum in potential energy surface exists for those nuclei. For an axially deformed nucleus with spatial reflection symmetry, the potentials in Equation (3) and densities in Equation (4) can be expanded in terms of the Legendre polynomials

$$f(\boldsymbol{r}) = \sum_\lambda f_\lambda(r) P_\lambda(\cos\theta), \qquad \lambda = 0, 2, 4, \cdots, \lambda_{\max} \tag{7}$$

with

$$f_\lambda(r) = \frac{2\lambda + 1}{4\pi} \int d\Omega f(\boldsymbol{r}) P_\lambda(\Omega). \tag{8}$$

In practical calculations, the deformed RHB Equation (1) is solved in a spherical Dirac Woods–Saxon basis, which is obtained by solving a Dirac equation with spherical scalar and vector potentials in Woods–Saxon forms [44]. Both the positive- and negative-energy states for the solution of Dirac equations are considered. In recent years, the completeness of the full Dirac space is crucial also for ab initio studies of nuclear structure [45] and nuclear matter [46,47]. The solution of the RHB equations provides us with the expansion coefficients of quasiparticle wave functions, from which new densities and potentials can be obtained. These quantities are iterated in the RHB equations until the convergence is achieved.

3. Numerical Details

The numerical details for constructing the DRHBc mass table have been examined thoroughly in refs. [33,37]. For nuclei with $8 \le Z \le 120$, according to ref. [37], the box size and the mesh size are $R_{\text{box}} = 20$ fm and $\Delta r = 0.1$ fm; the energy cutoff for the levels in the Fermi sea is $E_{\text{cut}}^+ = 300$ MeV; the number of states in the Dirac sea is taken to be the same as that in the Fermi sea; The angular momentum cutoff is $J_{\max} = 23/2\hbar$; the Legendre expansion truncation in Equation (7) is chosen as $\lambda_{\max} = 6,\ 8,\ 10$ for nuclei with $8 \le Z \le 70$, $72 \le Z \le 100$, and $102 \le Z \le 120$, respectively. For the pairing channel, the pairing strength $V_0 = -325.0$ MeV $\cdot$ fm^3 and the sharp pairing window of 100 MeV are used. A detailed examination of the numerical details for nuclei with $Z \geq 122$ is in progress. In this work, we will discuss the convergence of potential energy curves (PECs) with respect to the J_{max} in the region of superheavy nuclei. Other numerical details are aligned with the suggestions in Refs. [33,37], including typically the Legendre expansion truncation $\lambda_{\text{max}} = 10$ and the box size $R_{\text{box}} = 20$ fm.

The point-coupling density functional PC-PK1 is used in this work. Uncertainty quantification of DRHBc calculations from parameters can be obtained with thorough analysis of parameter fitting [48,49] and/or using machine learning techniques [50,51], which are beyond this scope and can be considered in future work. Regarding the extrapolation reliability of the theoretical framework and density functional adopted in this work, ref. [35] has shown that the DRHBc theory with PC-PK1 has an impressive performance.

In addition, for the following discussions, we temporarily adopt a systematic IUPAC name for nuclei with $Z > 100$ [52]. Specifically, nuclei with $Z = 134$ and $Z = 135$ are named by Utq and Utp, respectively.

4. Results and Discussion

Firstly, we investigate how the pattern of PECs changes as the step size of the initial quadrupole deformation parameter $\Delta\beta_2$ is varied. Here, β_2 has a standard meaning of deformations of the ellipsoid-like density distributions. One can easily expect that for too small $\Delta\beta_2$ the calculation costs cannot be ignored, while for too large $\Delta\beta_2$ the PECs are not smooth though to locate the local minimum. In Figure 1, we show the PECs of $^{384}_{134}\text{Utq}_{250}$ with three different initial deformation step sizes, i.e., $\Delta\beta_2 = 0.1, 0.05$, and 0.01. For clarity, the values for $\Delta\beta_2 = 0.1$ and 0.05 have been shifted by adding up 20 and 10 MeV. Obviously, constrained calculations with $\Delta\beta_2 = 0.01$ lead to an overly dense PEC, which is not necessary since there are 200 times calculations in the range of $-1.0 \leq \beta_2 \leq 1.0$. On the other hand, $\Delta\beta_2 = 0.1$ provides a loose PEC which is dangerous for possible missing of local minimum. Choosing $\Delta\beta_2 = 0.05$ is a very appropriate compromise considering both the computational cost and smoothness, as can be seen in Figure 1. We also show the unconstrained results, which is consistent with the constrained calculations. In the discussions below, if there is no explicit statement, the PECs are obtained with an initial deformation step size $\Delta\beta_2 = 0.05$.

In Figure 1, one finds that the ground state results from the competition between two local minima. One has an oblate deformation with $\beta_2 \simeq -0.2$, the other one has a prolate deformation with $\beta_2 \simeq 0.42$, which is much larger than the deformation for most nuclei [53]. According to the conventional deformed shell model, a larger prolate deformation causes a greater downward shift in single-particle levels with high angular momentum. Besides, the PECs in Figure 1 do not show fission possibilities even for $\beta_2 = 1.0$, which is unusual for such a superheavy nucleus. These two facts imply that the present cutoff of the angular momentum $J_{\text{max}} = 23/2\hbar$ might not be enough to contain high-order orbits for the superheavy nuclei.

Figure 2 shows how the PECs evolve with the increasing J_{max}. The nucleus $^{388}_{134}\text{Utq}_{254}$ is chosen as an example because, for the normal cutoff $J_{\text{max}} = 23/2\hbar$, the ground state is located at $\beta_2 = 0.44$, which has a large prolate deformation that needs to be checked. As expected, with the increase in J_{max}, the total energies with extreme deformations decrease and a fission pattern is found. For deformation β_2 in the range of $[-0.3, 0.3]$, as highlighted with two pink vertical dashed lines, $J_{\text{max}} = 23/2\hbar$ has already provided a large enough cutoff. In contrast, for a larger deformation with $|\beta_2| > 0.3$, $J_{\text{max}} = 31/2\hbar$ is a better cutoff not only to obtain converged results but also to find the correct ground state. Notice that in these calculations, the pairing effects are neglected for simplicity as conducted in ref. [33]. Furthermore, increasing the Legendre expansion truncation λ_{max} from 10 to 12 does not alter the conclusion here.

The observations from the previous two figures inspire us the following three steps in the determination of the ground state:

- Choose the initial deformation step size $\Delta\beta_2 = 0.05$ with $J_{\max} = 23/2\hbar$ to obtain a smooth enough PEC with acceptable computational costs.
- Check the total energies for large deformation with $|\beta_2| > 0.3$ by calculations with $J_{\max} = 31/2\hbar$.
- Perform unconstrained calculations in the vicinity of local minima in PEC. The configuration with the lowest total energy is the ground state.

Following the above strategy, we study the total energies of $^{388}_{134}\mathrm{Utq}_{254}$ as a function of deformation and present the results in Figure 3. In this case, we consider the pairing through the Bogoliubov theory, where the pairing strengths are fixed at $V_0 = -325.0$ MeV $\cdot$ fm^3 for $J_{\max} = 23/2\hbar$ [33] and $V_0 = -300.0$ MeV $\cdot$ fm^3 for $J_{\max} = 31/2\hbar$. Both combinations can reproduce the experimental odd–even mass differences in Ca and Pb isotope chains as in ref. [33]. It should be mentioned that a weaker pairing strength is required for a higher angular momentum cutoff because the zero-range pairing force adopted in Equation (6) needs to be renormalized to the enlarged model space.

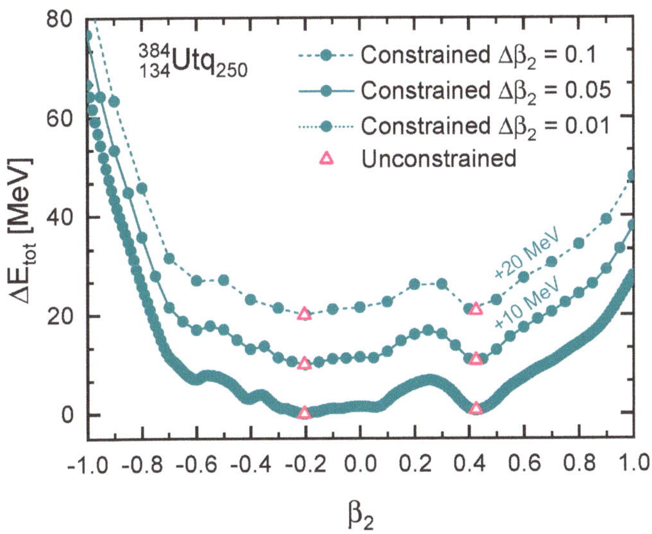

Figure 1. Potential energy curves of $^{384}_{134}\mathrm{Utq}_{250}$ in constrained DRHBc calculations with initial deformation step size $\Delta\beta_2 = 0.1$ (upper), 0.05 (middle), and 0.01 (lower). The unconstrained results are also shown with pink empty triangles. The energy for the ground state has been shifted to zero, while the values for $\Delta\beta_2 = 0.1$ and 0.05 have been additionally shifted by adding up 20 and 10 MeV, respectively.

From the PEC shown in Figure 3, the modification on total energy from $J_{\max} = 23/2\hbar$ to $J_{\max} = 31/2\hbar$ becomes more evident for larger deformation. Moreover, it is evident that there is competition between the two minima. This indicates that $\Delta\beta_2 = 0.05$ indeed provides a PEC with smoothness and clarity. Since the right minimum is around $\beta = 0.42$, it is necessary to perform examinations with a larger cutoff of angular momentum. By increasing $J_{\max}$ to $31/2\hbar$, the total energies around the prolate minimum decrease by 2–3 MeV, prohibiting a more advantageous stability in comparison with its competitor with oblate deformation. For the oblate side, we also perform calculations with $J_{\max} = 31/2\hbar$, while no new minimum is found in addition to the one with $\beta = -0.2$. After the unconstrained calculations, we come to the conclusion that the ground-state energy of $^{388}_{134}\mathrm{Utq}_{254}$ is -2516.952 MeV with deformation $\beta_2 = 0.461$.

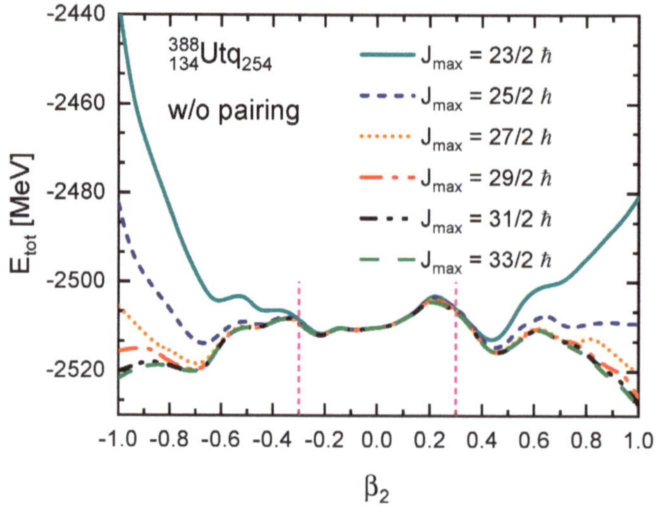

Figure 2. Potential energy curves of $^{388}_{134}$Utq$_{254}$ with the angular momentum cutoff J_{max} ranging from $23/2\hbar$ to $33/2\hbar$. The pairing correlation is neglected. Two pink vertical dashed lines at $|\beta_2| = 0.3$ are used to guide the eye.

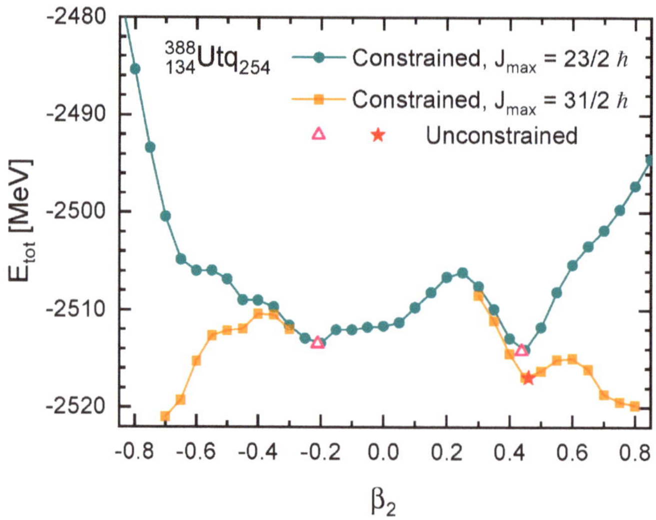

Figure 3. Potential energy curves of $^{388}_{134}$Utq$_{254}$ in constrained DRHBc calculations with the angular momentum cutoff $J_{max} = 23/2\hbar$ and initial deformation step size $\Delta\beta_2 = 0.05$. The results for constrained calculations with higher $J_{max} = 31/2\hbar$ for $|\beta_2| \geq 0.3$ and unconstrained calculations are also shown.

The strategy for determining the ground state is applied to three additional nuclei in the Utq ($Z = 134$) isotope chain with $N = 218$, 288, and 320. These three nuclei are chosen as representatives considering that the proton and neutron drip lines in this isotope chain are at $N = 202$ and $N = 350$, respectively. As depicted in Figure 4, $\Delta\beta_2 = 0.05$ is a good choice to obtain smooth and clear PECs, for both $J_{max} = 23/2\hbar$ and $J_{max} = 31/2\hbar$. For $N = 218$, the candidate for ground state at the oblate side has a deformation of $\beta_2 = -0.5$, while the candidate at the other side is less deformed. By increasing the cutoff of angular momentum, the ground state is confirmed with an oblate deformation with a lower total energy. For $N = 288$, the situation is reversed and it is the prolate candidate that has large deformation and needs to be checked. Interestingly, calculations with $J_{\text{max}} = 31/2\hbar$ find a more stable minimum with larger oblate deformation $\beta_2 = -0.563$, which does not show up with $J_{\text{max}} = 23/2\hbar$. This indicates that the examination with a larger cutoff is of high necessity even for cases where the oblate and/or prolate deformation of ground-state candidates is not larger than 0.3. For $N = 320$ shown in the lower panel in Figure 4, the PEC is rather simple with $J_{\text{max}} = 23/2\hbar$, showing a softness around the spherical configuration. However, with $J_{\text{max}} = 31/2\hbar$, an oblate minimum shows up, which is found to be the ground state.

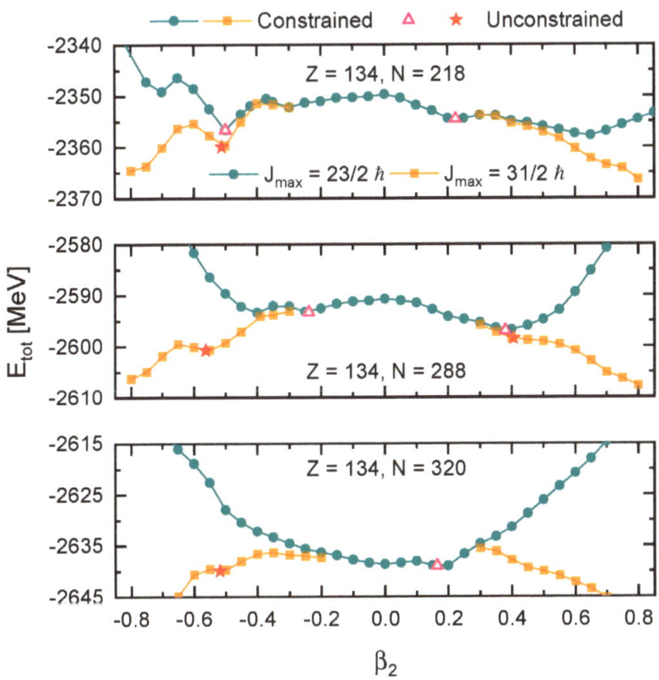

Figure 4. Potential energy curves of Utq nucleus ($Z = 134$) with $N = 218$ (**upper**), 288 (**middle**), and 320 (**lower**) in constrained DRHBc calculations with initial deformation step size $\Delta\beta_2 = 0.1$. Results with $J_{\text{max}} = 23/2\hbar$, $|\beta_2| \leq 0.8$ as well as the ones with $J_{\text{max}} = 31/2\hbar$, $|\beta_2| \geq 0.3$ for $N = 218$, 288, and 320 are shown. The unconstrained results are also shown with pink empty triangles and red stars.

All the aforementioned discussions are for even–even nuclei. The three steps are further applied to odd-A nuclei, where the time-reversal invariance is retained by blocking the quasiparticle configurations within the equal-filling approximation [37]. In Figure 5, the PECs of three nuclei in the Utp ($Z = 135$) isotope chain with N = 218, 288 and 320 are given. Despite the difficulties in a quite demanding computational procedure, a smooth and clear PEC can be obtained with initial deformation step size $\Delta\beta_2 = 0.05$. Besides, by increasing J_{max} from $23/2\hbar$ to $31/2\hbar$, one can identify the correct ground state, no matter if it is oblate or prolate with a deformation parameter $|\beta|$ smaller or larger than 0.3.

In Table 1, we tabulate the ground-state properties of several nuclei shown from Figures 3–5, including the binding energy $E_{\text{b}}^{\text{cal}} \equiv -E_{\text{tot}}$ from the DRHBc calculations, the binding energy plus rotational correction energy $E_{\text{b+rot}}^{\text{cal}}$, the binding energy per nucleon $E_{\text{b+rot}}^{\text{cal}}/A$, and the quadrupole deformation β_2.

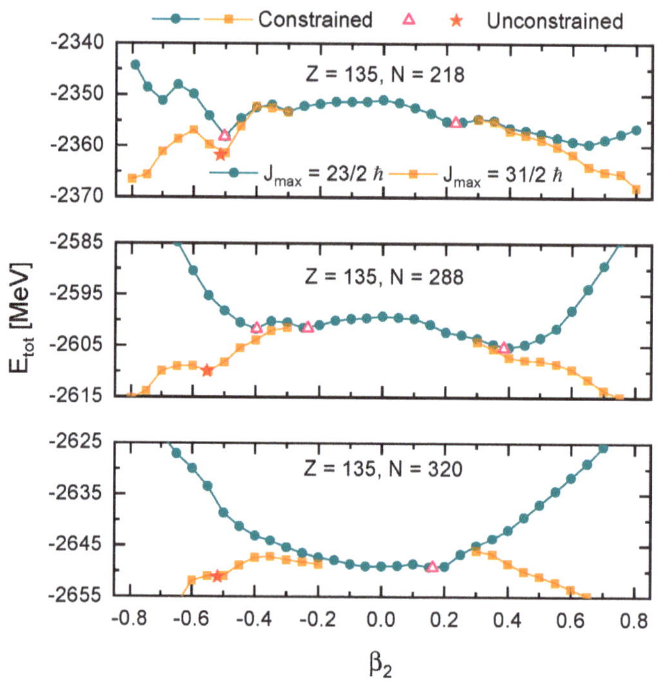

Figure 5. Similar to Figure 4, but for Utp nucleus ($Z = 135$).

Table 1. Ground-state properties of several nuclei from Figures 3–5 calculated by the DRHBc theory.

A	N	E_b^{cal} (MeV)	E_{b+rot}^{cal} (MeV)	E_{b+rot}^{cal}/A (MeV)	β_2
$Z = 134$ (Utq)					
352	218	2360.01	2363.23	6.714	−0.511
388	254	2516.95	2518.84	6.492	0.461
422	288	2600.77	2603.71	6.170	−0.563
454	320	2639.96	2642.49	5.821	−0.518
$Z = 135$ (Utp)					
353	218	2361.76	2364.88	6.700	−0.515
423	288	2609.93	2612.71	6.177	−0.554
455	320	2651.20	2653.72	5.832	−0.529

5. Summary

Determining the ground states of superheavy nuclei is particularly challenging but important for both experimental and theoretical studies. Starting from the PC-PK1 point-coupling density functional, we studied the properties of superheavy nuclei with $Z = 134$ and 135 within the deformed relativistic Hartree–Bogoliubov theory in continuum (DRHBc). By investigating how the pattern of potential energy curves (PECs) varies with the changes in the step size for the initial quadrupole deformation parameter $\Delta\beta_2$ and its convergence with respect to the cutoff of angular momentum J_{max}, we propose three steps for determining the ground state, in balancing the computational complexity and accuracy:

- Choose the initial deformation step size $\Delta\beta_2 = 0.05$ with $J_{max} = 23/2\hbar$ to obtain a smooth enough PEC with acceptable computational costs.
- Check the total energies for large deformation with $|\beta_2| > 0.3$ by calculations with $J_{max} = 31/2\hbar$.

- Perform unconstrained calculations in the vicinity of local minima in PEC. The configuration with the lowest total energy is the ground state.

These three steps are then applied to three representative nuclei in the $Z = 134$ isotope chain with $N = 218$, 288, and 320. The ground states for the first two nuclei arise from the competition between oblate and prolate configurations which are already observed with $J_{\text{max}} = 23/2\hbar$, while the ground state for $N = 320$ is obtained only after performing calculations with $J_{\text{max}} = 31/2\hbar$. The validation of the strategy is also confirmed for odd-A nuclei with $Z = 135$ and $N = 218$, 288 and 320. This work provides guidance for large-scale calculations of superheavy nuclei as a new extension of the DRHBc mass table.

Author Contributions: Conceptualization, S.W.; methodology, P.G. and C.P.; software, P.G. and C.P.; formal analysis, S.W.; writing—original draft preparation, S.W.; writing—review and editing, All authors; visualization, S.W.; project administration, S.W.; funding acquisition, S.W. All authors have read and agreed to the published version of the manuscript.

Funding: This research was funded in part by the National Natural Science Foundation of China (NSFC) under Grant No. 12205030, and the Project No. 2024CDJXY022 supported by the Fundamental Research Funds for the Central Universities. Part of this work was achieved by using the supercomputer OCTOPUS at the Cybermedia Center, Osaka University under the support of Research Center for Nuclear Physics of Osaka University.

Data Availability Statement: The raw data supporting the conclusions of this article will be made available by the authors on request.

Acknowledgments: S.W. thanks Xin-Hui Wu for fruitful discussions. Helpful discussions with members of the DRHBc Mass Table Collaboration are highly appreciated.

Conflicts of Interest: The authors declare no conflicts of interest.

References

1. Lunney, D.; Pearson, J.M.; Thibault, C. Recent trends in the determination of nuclear masses. *Rev. Mod. Phys.* **2003**, *75*, 1021–1082. [CrossRef]
2. Blaum, K. High-accuracy mass spectrometry with stored ions. *Phys. Rep.* **2006**, *425*, 1–78. [CrossRef]
3. Mumpower, M.; Surman, R.; McLaughlin, G.; Aprahamian, A. The impact of individual nuclear properties on r-process nucleosynthesis. *Prog. Part. Nucl. Phys.* **2016**, *86*, 86–126. [CrossRef]
4. Jiang, X.F.; Wu, X.H.; Zhao, P.W. Sensitivity Study of r-process Abundances to Nuclear Masses. *Astrophys. J.* **2021**, *915*, 29. [CrossRef]
5. Baldo, M.; Burgio, G.F.; Schulze, H.J.; Taranto, G. Nucleon effective masses within the Brueckner-Hartree-Fock theory: Impact on stellar neutrino emission. *Phys. Rev. C* **2014**, *89*, 048801. [CrossRef]
6. Hofmann, S.; Münzenberg, G. The discovery of the heaviest elements. *Rev. Mod. Phys.* **2000**, *72*, 733–767. [CrossRef]
7. Oganessian, Y.T.; Utyonkov, V.K. Super-heavy element research. *Rep. Prog. Phys.* **2015**, *78*, 036301. [CrossRef] [PubMed]
8. Giuliani, S.A.; Matheson, Z.; Nazarewicz, W.; Olsen, E.; Reinhard, P.G.; Sadhukhan, J.; Schuetrumpf, B.; Schunck, N.; Schwerdtfeger, P. Colloquium: Superheavy elements: Oganesson and beyond. *Rev. Mod. Phys.* **2019**, *91*, 011001. [CrossRef]
9. Smits, O.R.; Düllmann, C.E.; Indelicato, P.; Nazarewicz, W.; Schwerdtfeger, P. The quest for superheavy elements and the limit of the periodic table. *Nat. Rev. Phys.* **2024**, *6*, 86–98. [CrossRef]
10. Gregorich, K. Simulation of recoil trajectories in gas-filled magnetic separators. *Nucl. Instrum. Methods Phys. Res. Sect. A Accel. Spectrometers Detect. Assoc. Equip.* **2013**, *711*, 47–59. [CrossRef]
11. Zhang, Z.Y.; Gan, Z.G.; Yang, H.B.; Ma, L.; Huang, M.H.; Yang, C.L.; Zhang, M.M.; Tian, Y.L.; Wang, Y.S.; Sun, M.D.; et al. New Isotope ^{220}Np: Probing the Robustness of the $N = 126$ Shell Closure in Neptunium. *Phys. Rev. Lett.* **2019**, *122*, 192503. [CrossRef] [PubMed]
12. Ma, L.; Zhang, Z.Y.; Gan, Z.G.; Zhou, X.H.; Yang, H.B.; Huang, M.H.; Yang, C.L.; Zhang, M.M.; Tian, Y.L.; Wang, Y.S.; et al. Short-Lived α-Emitting Isotope ^{222}Np and the Stability of the $N = 126$ Magic Shell. *Phys. Rev. Lett.* **2020**, *125*, 032502. [CrossRef] [PubMed]
13. Myers, W.D.; Swiatecki, W.J. Nuclear masses and deformations. *Nucl. Phys.* **1966**, *81*, 1–60. [CrossRef]
14. Viola, V.; Seaborg, G. Nuclear systematics of the heavy elements—II Lifetimes for alpha, beta and spontaneous fission decay. *J. Inorg. Nucl. Chem.* **1966**, *28*, 741–761. [CrossRef]
15. Oganessian, Y.T.; Abdullin, F.S.; Alexander, C.; Binder, J.; Boll, R.A.; Dmitriev, S.N.; Ezold, J.; Felker, K.; Gostic, J.M.; Grzywacz, R.K.; et al. Production and Decay of the Heaviest Nuclei ^{293,294}117 and ^{294}118. *Phys. Rev. Lett.* **2012**, *109*, 162501. [CrossRef]
16. Aboussir, Y.; Pearson, J.; Dutta, A.; Tondeur, F. Nuclear mass formula via an approximation to the Hartree–Fock method. *At. Data Nucl. Data Tables* **1995**, *61*, 127–176. [CrossRef]

17. Wang, N.; Liu, M.; Wu, X.; Meng, J. Surface diffuseness correction in global mass formula. *Phys. Lett. B* **2014**, *734*, 215–219. [CrossRef]
18. Möller, P.; Sierk, A.; Ichikawa, T.; Sagawa, H. Nuclear ground-state masses and deformations: FRDM(2012). *At. Data Nucl. Data Tables* **2016**, *109–110*, 1–204. [CrossRef]
19. Samyn, M.; Goriely, S.; Heenen, P.H.; Pearson, J.; Tondeur, F. A Hartree–Fock–Bogoliubov mass formula. *Nucl. Phys. A* **2002**, *700*, 142–156. [CrossRef]
20. Goriely, S.; Chamel, N.; Pearson, J.M. Skyrme-Hartree-Fock-Bogoliubov Nuclear Mass Formulas: Crossing the 0.6 MeV Accuracy Threshold with Microscopically Deduced Pairing. *Phys. Rev. Lett.* **2009**, *102*, 152503. [CrossRef] [PubMed]
21. Goriely, S.; Hilaire, S.; Girod, M.; Péru, S. First Gogny-Hartree-Fock-Bogoliubov Nuclear Mass Model. *Phys. Rev. Lett.* **2009**, *102*, 242501. [CrossRef]
22. Lalazissis, G.; Raman, S.; Ring, P. Ground-state properties of even–even nuclei in the relativistic mean-field theory. *At. Data Nucl. Data Tables* **1999**, *71*, 1–40. [CrossRef]
23. Geng, L.; Toki, H.; Meng, J. Masses, Deformations and Charge Radii—Nuclear Ground-State Properties in the Relativistic Mean Field Model. *Prog. Theor. Phys.* **2005**, *113*, 785–800. [CrossRef]
24. Afanasjev, A.; Agbemava, S.; Ray, D.; Ring, P. Nuclear landscape in covariant density functional theory. *Phys. Lett. B* **2013**, *726*, 680–684. [CrossRef]
25. Agbemava, S.E.; Afanasjev, A.V.; Ray, D.; Ring, P. Global performance of covariant energy density functionals: Ground state observables of even–even nuclei and the estimate of theoretical uncertainties. *Phys. Rev. C* **2014**, *89*, 054320. [CrossRef]
26. Afanasjev, A.V.; Agbemava, S.E.; Ray, D.; Ring, P. Neutron drip line: Single-particle degrees of freedom and pairing properties as sources of theoretical uncertainties. *Phys. Rev. C* **2015**, *91*, 014324. [CrossRef]
27. Xia, X.; Lim, Y.; Zhao, P.; Liang, H.; Qu, X.; Chen, Y.; Liu, H.; Zhang, L.; Zhang, S.; Kim, Y.; et al. The limits of the nuclear landscape explored by the relativistic continuum Hartree–Bogoliubov theory. *At. Data Nucl. Data Tables* **2018**, *121–122*, 1–215. [CrossRef]
28. Zhang, K.; Cheoun, M.K.; Choi, Y.B.; Chong, P.S.; Dong, J.; Dong, Z.; Du, X.; Geng, L.; Ha, E.; He, X.T.; et al. Nuclear mass table in deformed relativistic Hartree–Bogoliubov theory in continuum, I: Even–even nuclei. *At. Data Nucl. Data Tables* **2022**, *144*, 101488. [CrossRef]
29. Guo, P.; Cao, X.; Chen, K.; Chen, Z.; Cheoun, M.K.; Choi, Y.B.; Lam, P.C.; Deng, W.; Dong, J.; Du, P.; et al. Nuclear mass table in deformed relativistic Hartree–Bogoliubov theory in continuum, II: Even-Z nuclei. *At. Data Nucl. Data Tables* **2024**, *158*, 101661. [CrossRef]
30. Geng, J.; Long, W.H. Relativistic Hartree-Fock-Bogoliubov model for axially deformed nuclei. *Phys. Rev. C* **2022**, *105*, 034329. [CrossRef]
31. Geng, J.; Zhao, P.W.; Niu, Y.F.; Long, W.H. A coherent microscopic picture for the exotic structure of 11Be. *Phys. Lett. B* **2024**, *858*, 139036. [CrossRef]
32. Li, L.; Meng, J.; Ring, P.; Zhao, E.G.; Zhou, S.G. Deformed relativistic Hartree–Bogoliubov theory in continuum. *Phys. Rev. C* **2012**, *85*, 024312. [CrossRef]
33. Zhang, K.; Cheoun, M.K.; Choi, Y.B.; Chong, P.S.; Dong, J.; Geng, L.; Ha, E.; He, X.; Heo, C.; Ho, M.C.; et al. Deformed relativistic Hartree–Bogoliubov theory in continuum with a point-coupling functional: Examples of even–even Nd isotopes. *Phys. Rev. C* **2020**, *102*, 024314. [CrossRef]
34. Zhao, P.W.; Li, Z.P.; Yao, J.M.; Meng, J. New parametrization for the nuclear covariant energy density functional with a point-coupling interaction. *Phys. Rev. C* **2010**, *82*, 054319. [CrossRef]
35. Zhang, K.; He, X.; Meng, J.; Pan, C.; Shen, C.; Wang, C.; Zhang, S. Predictive power for superheavy nuclear mass and possible stability beyond the neutron drip line in deformed relativistic Hartree–Bogoliubov theory in continuum. *Phys. Rev. C* **2021**, *104*, L021301. [CrossRef]
36. DRHBc Mass Table Collaboration. Available online: http://drhbctable.jcnp.org (accessed on 20 December 2024).
37. Pan, C.; Cheoun, M.K.; Choi, Y.B.; Dong, J.; Du, X.; Fan, X.H.; Gao, W.; Geng, L.; Ha, E.; He, X.T.; et al. Deformed relativistic Hartree–Bogoliubov theory in continuum with a point-coupling functional. II. Examples of odd Nd isotopes. *Phys. Rev. C* **2022**, *106*, 014316. [CrossRef]
38. Dechargé, J.; Berger, J.F.; Dietrich, K.; Weiss, M. Superheavy and hyperheavy nuclei in the form of bubbles or semi-bubbles. *Phys. Lett. B* **1999**, *451*, 275–282. [CrossRef]
39. Bender, M.; Nazarewicz, W.; Reinhard, P.G. Shell stabilization of super- and hyperheavy nuclei without magic gaps. *Phys. Lett. B* **2001**, *515*, 42–48. [CrossRef]
40. Afanasjev, A.; Agbemava, S.; Gyawali, A. Hyperheavy nuclei: Existence and stability. *Phys. Lett. B* **2018**, *782*, 533–540. [CrossRef]
41. Zhou, S.G.; Meng, J.; Ring, P.; Zhao, E.G. Neutron halo in deformed nuclei. *Phys. Rev. C* **2010**, *82*, 011301. [CrossRef]
42. Zhang, W.; Meng, J.; Zhang, S.; Geng, L.; Toki, H. Magic numbers for superheavy nuclei in relativistic continuum Hartree–Bogoliubov theory. *Nucl. Phys. A* **2005**, *753*, 106–135. [CrossRef]
43. Brodziński, W.; Skalski, J. Predictions for superheavy elements beyond $Z = 126$. *Phys. Rev. C* **2013**, *88*, 044307. [CrossRef]
44. Zhou, S.G.; Meng, J.; Ring, P. Spherical relativistic Hartree theory in a Woods–Saxon basis. *Phys. Rev. C* **2003**, *68*, 034323. [CrossRef]

45. Shen, S.; Liang, H.; Long, W.; Meng, J.; Ring, P. Towards an ab initio covariant density functional for nuclear structure. *Prog. Part. Nucl. Phys.* **2019**, *109*, 103713. [CrossRef]
46. Wang, S.; Zhao, Q.; Ring, P.; Meng, J. Nuclear matter in relativistic Brueckner-Hartree-Fock theory with Bonn potential in the full Dirac space. *Phys. Rev. C* **2021**, *103*, 054319. [CrossRef]
47. Wang, S.; Tong, H.; Wang, C.; Zhao, Q.; Ring, P.; Meng, J. Tensor-force effects on nuclear matter in relativistic ab initio theory. *Sci. Bull.* **2024**, *69*, 2166–2169. [CrossRef] [PubMed]
48. Nikšić, T.; Paar, N.; Reinhard, P.G.; Vretenar, D. Optimizing relativistic energy density functionals: Covariance analysis. *J. Phys. G: Nucl. Part. Phys.* **2015**, *42*, 034008. [CrossRef]
49. Agbemava, S.E.; Afanasjev, A.V.; Taninah, A. Propagation of statistical uncertainties in covariant density functional theory: Ground state observables and single-particle properties. *Phys. Rev. C* **2019**, *99*, 014318. [CrossRef]
50. Wu, X.H.; Pan, C.; Zhang, K.Y.; Hu, J. Nuclear mass predictions of the relativistic continuum Hartree–Bogoliubov theory with the kernel ridge regression. *Phys. Rev. C* **2024**, *109*, 024310. [CrossRef]
51. Guo, Y.Y.; Yu, T.; Wu, X.H.; Pan, C.; Zhang, K.Y. Nuclear mass predictions of the relativistic continuum Hartree–Bogoliubov theory with the kernel ridge regression. II. Odd-even effects. *Phys. Rev. C* **2024**, *110*, 064310. [CrossRef]
52. Chat, J. Recommendations for the Naming of Elements of Atomic Numbers Greater than 100. *Pure Appl. Chem.* **1979**, *51*, 381–384. [CrossRef]
53. Yang, Y.L.; Wang, Y.K.; Zhao, P.W.; Li, Z.P. Nuclear landscape in a mapped collective Hamiltonian from covariant density functional theory. *Phys. Rev. C* **2021**, *104*, 054312. [CrossRef]

particles

MDPI

Article

Examination of Possible Proton Magic Number $Z = 126$ with the Deformed Relativistic Hartree-Bogoliubov Theory in Continuum

Cong Pan [1,2,*] and Xin-Hui Wu [3]

1 Department of Physics, Anhui Normal University, Wuhu 241000, China
2 Center for Exotic Nuclear Studies, Institute for Basic Science, Daejeon 34126, Republic of Korea
3 Department of Physics, Fuzhou University, Fuzhou 350108, China; wuxinhui@fzu.edu.cn
* Correspondence: cpan@ahnu.edu.cn

Abstract: Whether $Z = 126$ is a proton magic number has been controversial in nuclear physics. The even-even $_{126}$Ubh isotopes are calculated based on the DRHBc calculations with PC-PK1. The evolutions of quadrupole deformation and pairing energies for neutron and proton are analyzed to study the possible nuclear magicity. Spherical shape occurs and neutron pairing energy vanishes at $N = 258$ and 350, which are the results of possible neutron magicity, while the proton pairing energy never vanishes in Ubh isotopes, which does not support the proton magicity at $Z = 126$. In the single-proton spectrum, there is no discernible gap at $Z = 126$, while significant gaps appear at $Z = 120$ and 138. Therefore, $Z = 126$ is not supported as a proton magic number, while $Z = 120$ and 138 are suggested as candidates of proton magic numbers.

Keywords: superheavy nuclei; magic number; deformation; shell structure

1. Introduction

The exploration of the limit of nuclear existence has been a very fascinating topic in nuclear physics [1–4]. With neutron number N and proton number Z as horizontal and vertical axes, respectively, the southwest coast of the nuclear landscape corresponds the beginning of periodic table, H, the lightest element. The northwest and southeast coasts are the proton and neutron drip lines, respectively. The proton drip line has been experimentally determined up to Np with $Z = 93$ [5], while the neutron drip line has only been determined up to Ne with $Z = 10$ [6]. Near the drip lines, many interesting phenomena are discovered, including the nuclear halo [7], changes of nuclear magic numbers [8] and pygmy resonances [9], and have attracted worldwide attentions. The northeast coast of the nuclear landscape corresponds to the upper limit of mass number A, and is still unknown because data are extremely limited. The heaviest nuclides discovered so far are ^{294}Og and ^{294}Ts [10], with nucleon numbers $(Z, N) = (118, 176)$ and $(117, 177)$, respectively. These two nuclides have not reached the so-called "island of stability", which is theoretically predicted based on the possible neutron magic number $N = 184$ and proton magic number $Z = 114$ [11–14].

The concept of nuclear magic numbers refers to the particular stability of nuclei with certain nucleon numbers. Such stability is reflected on the extra nuclear binding energy, and a more intuitive manifestation is a significantly larger separation energy than that of the next nuclide. The nuclear magicity is closely related to the nuclear shell effect, which is a hallmark characteristic in the atomic nucleus as a quantum system. The experimentally confirmed magic numbers are the neutron numbers $N = 2, 8, 20, 28, 50, 82, 126$ and proton numbers $Z = 2, 8, 20, 28, 50, 82$ [15]. These magic numbers can be perfectly reproduced by assuming a harmonic oscillator potential plus spin-orbital coupling [16,17]. In a naïve consideration with this model, extra stability is suggested at nucleon numbers $2, 8, 20, 28, 50, 82, 126, 184, 258, 350, \ldots$, which not only explains all the magic numbers above, but also provides a reference for new possible magic numbers such as 184 and 258, as well as $Z = 126$. The traditional concept of magic number has been implicitly associated with spherical symmetry. Recent development has extended this concept to exotic shape symmetries and fourfold octupole magic number $N = 136$ and 196 have been predicted with symmetry induced increase in stability [18,19]. The nuclei with exotic shape are predicted to form islands in the superheavy region [20].

The possibility of a proton magic number at $Z = 126$ has been discussed since as early as 1955 [21]. Following the establishment of the nuclear shell model, the potential existence of stable nuclides near $Z = 126$ was studied and there were also controversies in the conclusions [22–25]. In addition to $Z = 126$, subsequent efforts have also been made to study other possible magic numbers in the superheavy nuclear region. Based on phenomenological models such as finite-range droplet model (FRDM), $Z = 114$ and $N = 184$ are predicted to have large shell gaps, corresponding to new magic numbers [26]. In the FRDM, other large shell gaps are predicted to be located at $N = 162$ and $Z = 104, 106, 108, 110$, instead of $Z = 126$. In order to predict experimentally unknown areas, the prediction stability is a crucial issue to address. The predictive stability of micro-macro models has been investigated thoroughly in many literatures, such as Refs. [27,28] which use the inverse problem theory of applied mathematics and Monte Carlo simulations. For microscopic models, the theoretical uncertainties have also been investigated in a systematic manner [29,30].

The relativistic density functional theory has been proven to be a powerful tool in nuclear physics, due to its successful descriptions on many nuclear phenomena [4]. For the exotic nuclei far away from stability line, the occupation of single-nucleon spectrum is very close to the continuum threshold, and the pairing interaction can scatter nucleons from bound states to resonant states in the continuum, leading to a more diffuse density and the dripline locations might be influenced, called the continuum effects [31–34]. A proper treatment for continuum is solving the nucleon system described by Bogoliubov transformation in coordinate space, where wave functions are approximated on a spatial lattice, and the continuum is discretized by suitably large box boundary conditions [35–38]. Based on the relativistic density functional theory and solving the problem in coordinate space, the spherical relativistic continuum Hartree-Bogoliubov (RCHB) theory was developed [38,39], which can properly take into account the effects of pairing correlations and continuum for the nuclei near the limit of nuclear landscape. The RCHB theory has been applied to many studies for both stable and exotic nuclei, including describing the halo in ^{11}Li [39], predicting giant halos [40], interpreting the pseudospin symmetry in exotic nuclei [41], reproducing the interaction cross section and charge-changing cross sections in light nuclei [42,43], etc. Based on the RCHB theory, the shell structures for superheavy nuclei were studied, and $Z = 120, 132, 138$, and $N = 172, 184, 198, 228, 238, 258$ were suggested to be the magic numbers [44]. In Ref. [45], the first nuclear mass table for the nuclei with $8 \leqslant Z \leqslant 120$ that incorporates continuum effects was constructed based on the RCHB theory, where the evolution of shell structures and magic numbers were studied.

Considering that most nuclei deviate from the spherical shape, the deformed relativistic Hartree-Bogoliubov theory in continuum (DRHBc) was developed [46,47]. The DRHBc theory takes into account the effects of deformation, pairing correlation and continuum, and can provide proper descriptions for both the stable nuclei and the unstable exotic nuclei near the boundary of nuclear landscape. As the advantages of the RCHB theory are inherited and the deformation degrees of freedom are further included, the DRHBc theory has been applied in many studies on exotic nuclei, including the halo structures [48–56], dripline locations [57–59], evolution of deformation and shape coexistence [60–63], etc. Recently, a nuclear mass table for the nuclei with $8 \leqslant Z \leqslant 120$ is in progress [64,65], and the even-Z part has been established [66,67]. The mass description for superheavy nuclei reach accuracy of several hundred keV [57] in reproducing available experimental data as well as empirical data in AME2020 [68]. Based on the DRHBc theory, $Z = 120$ is suggested as a candidate of proton magic number [66], while $Z = 126$ has not been considered yet.

In this work, the DRHBc theory is employed to examine the possible proton magic number $Z = 126$. The $Z = 126$ element has a temporary systematic IUPAC name as Unbiohexium (Ubh) [69]. The theoretical framework is briefly introduced in Section 2. The numerical details are given in Section 3. The results and discussions are presented in Section 4. Finally, a summary is given in Section 5.

2. Theoretical Framework

The starting point of the relativistic density functional theory is an effective Lagrangian density with either the meson-exchange or point-coupling interactions [4]. For the point-coupling interaction, the Lagrangian density reads

$$\begin{aligned}
\mathcal{L} =&\bar{\psi}(i\gamma_\mu\partial^\mu - M)\psi \\
&-\frac{1}{2}\alpha_S(\bar{\psi}\psi)(\bar{\psi}\psi) - \frac{1}{2}\alpha_V(\bar{\psi}\gamma_\mu\psi)(\bar{\psi}\gamma^\mu\psi) - \frac{1}{2}\alpha_{TV}(\bar{\psi}\vec{\tau}\gamma_\mu\psi)(\bar{\psi}\vec{\tau}\gamma^\mu\psi) \\
&-\frac{1}{2}\alpha_{TS}(\bar{\psi}\vec{\tau}\psi)(\bar{\psi}\vec{\tau}\psi) - \frac{1}{3}\beta_S(\bar{\psi}\psi)^3 - \frac{1}{4}\gamma_S(\bar{\psi}\psi)^4 - \frac{1}{4}\gamma_V[(\bar{\psi}\gamma_\mu\psi)(\bar{\psi}\gamma^\mu\psi)]^2 \\
&-\frac{1}{2}\delta_S\partial_\nu(\bar{\psi}\psi)\partial^\nu(\bar{\psi}\psi) - \frac{1}{2}\delta_V\partial_\nu(\bar{\psi}\gamma_\mu\psi)\partial^\nu(\bar{\psi}\gamma^\mu\psi) - \frac{1}{2}\delta_{TV}\partial_\nu(\bar{\psi}\vec{\tau}\gamma_\mu\psi)\partial^\nu(\bar{\psi}\vec{\tau}\gamma^\mu\psi) \\
&-\frac{1}{2}\delta_{TS}\partial_\nu(\bar{\psi}\vec{\tau}\psi)\partial^\nu(\bar{\psi}\vec{\tau}\psi) - \frac{1}{4}F^{\mu\nu}F_{\mu\nu} - e\bar{\psi}\gamma^\mu\frac{1-\tau_3}{2}A_\mu\psi,
\end{aligned} \tag{1}$$

where M is the nucleon mass, e is the charge unit, and ψ is the field operator for nucleon. The constants $\alpha, \beta, \gamma, \delta$ terms correspond to four-fermion, third-order, fourth-order, and derivative couplings terms, respectively [70,71]. The subscripts S, V, and T mean scalar, vector, and isovector, respectively. The last two terms describe the electromagnetic interaction in the standard quantum electrodynamics Lagrangian, and A_μ and $F_{\mu\nu}$ are the four-vector potential and field strength tensor of the electromagnetic field, respectively.

Starting from Equation (1) and performing Legendre transformation, the Hamiltonian is obtained, which, by using the mean-field approximation, is then utilized to derive the energy density functional. By simultaneously treating the mean fields and pairing correlations in a self-consistent manner, the nucleons are described by the the relativistic Hartree-Bogoliubov (RHB) equation [72],

$$\begin{pmatrix} \hat{h}_D - \lambda_\tau & \hat{\Delta} \\ -\hat{\Delta}^* & -\hat{h}_D^* + \lambda_\tau \end{pmatrix} \begin{pmatrix} U_k \\ V_k \end{pmatrix} = E_k \begin{pmatrix} U_k \\ V_k \end{pmatrix}, \tag{2}$$

where λ_τ is the Fermi energy of neutron or proton ($\tau = n, p$), E_k is the quasiparticle energy, U_k and V_k are the quasiparticle wave functions, $\hat{h}_D$ is the Dirac Hamiltonian, and $\hat{\Delta}$ is the pairing potential.

In the coordinate space, the Dirac Hamiltonian can be written as

$$h_D(\boldsymbol{r}) = \boldsymbol{\alpha} \cdot \boldsymbol{p} + V(\boldsymbol{r}) + \beta[M + S(\boldsymbol{r})], \tag{3}$$

where $S(\boldsymbol{r})$ and $V(\boldsymbol{r})$ are scalar and vector potentials, respectively,

$$S(\boldsymbol{r}) = \alpha_S\rho_S + \beta_S\rho_S^2 + \gamma_S\rho_S^3 + \delta_S\triangle\rho_S, \tag{4}$$

$$V(\boldsymbol{r}) = \alpha_V\rho_V + \gamma_V\rho_V^3 + \delta_V\triangle\rho_V + eA^0 + \alpha_{TV}\tau_3\rho_3 + \delta_{TV}\tau_3\triangle\rho_3. \tag{5}$$

ρ_S, ρ_V, and ρ_3 are the local densities, and are defined as

$$\rho_S(\boldsymbol{r}) = \sum_{k>0} V_k^\dagger(\boldsymbol{r})\gamma_0 V_k(\boldsymbol{r}), \tag{6}$$

$$\rho_V(\boldsymbol{r}) = \sum_{k>0} V_k^\dagger(\boldsymbol{r}) V_k(\boldsymbol{r}), \tag{7}$$

$$\rho_3(\boldsymbol{r}) = \sum_{k>0} V_k^\dagger(\boldsymbol{r})\tau_3 V_k(\boldsymbol{r}). \tag{8}$$

$k > 0$ means that the summation runs over the quasiparticle states in the Fermi sea, which corresponds to the so-called no-sea approximation.

The pairing potential $\hat{\Delta}$ reads

$$\Delta(\boldsymbol{r}_1, \boldsymbol{r}_2) = V^{pp}(\boldsymbol{r}_1, \boldsymbol{r}_2)\kappa(\boldsymbol{r}_1, \boldsymbol{r}_2), \tag{9}$$

where V^{pp} is the pairing force, and κ is the pairing tensor. In this work, the density-dependent zero-range pairing force

$$V^{pp}(\boldsymbol{1}_1, \boldsymbol{r}_2) = V_0 \frac{1}{2}(1 - P^\sigma)\delta(\boldsymbol{r}_1 - \boldsymbol{r}_2)\left(1 - \frac{\rho(\boldsymbol{r}_1)}{\rho_{\text{sat}}}\right) \tag{10}$$

is adopted. The pairing tensor is defined by using quasiparticle wavefunctions as [73]

$$\kappa = U * V^T. \tag{11}$$

In the DRHBc theory, since axial deformation and spatial reflection symmetry are assumed, the potentials and densities are expanded in terms of the Legendre polynomials,

$$f(\boldsymbol{r}) = \sum_\lambda f_\lambda(r) P_\lambda(\cos\theta), \qquad \lambda = 0, 2, 4, \ldots, \lambda_{\text{max}}. \tag{12}$$

It is worth mentioning that this work focuses on axial and spatial reflection symmetries, while higher-order deformation, spatial-reflection asymmetry as well as triaxial shapes, which might also play important roles in heavy nuclei [19,20,28,30], are not included here. The conclusion is limited due to such an assumption on symmetry, and future works are expected by taking into account the full degrees of freedom that nuclei can take.

In order to correctly take into account the continuum effect, the deformed RHB Equation (2) are solved in a spherical Dirac Woods-Saxon basis [74], which can properly describe the asymptotic behavior of the density distribution at a large r.

After self-consistently solving the RHB equations, the expectation values can be calculated. The total energy is calculated as [47,64]

$$\begin{aligned} E_{\text{tot}} = &\sum_{k>0}(\lambda_\tau - E_k)v_k^2 - E_{\text{pair}} \\ &- \int d^3\boldsymbol{r}\left(\frac{1}{2}\alpha_S\rho_S^2 + \frac{1}{2}\alpha_V\rho_V^2 + \frac{1}{2}\alpha_{TV}\rho_3^2 + \frac{2}{3}\beta_S\rho_S^3 + \frac{3}{4}\gamma_S\rho_S^4 + \frac{3}{4}\gamma_V\rho_V^4\right. \\ &\left. + \frac{1}{2}\delta_S\rho_S\triangle\rho_S + \frac{1}{2}\delta_V\rho_V\triangle\rho_V + \frac{1}{2}\delta_{TV}\rho_3\triangle\rho_3 + \frac{1}{2}\rho_p e A^0\right) + E_{\text{c.m.}}, \end{aligned} \tag{13}$$

where

$$v_k^2 = \int d^3\boldsymbol{r} V_k^\dagger(\boldsymbol{r}) V_k(\boldsymbol{r}), \tag{14}$$

and E_{pair} and $E_{\text{c.m.}}$ are the pairing energy and center-of-mass correction energy [47], respectively. It should be mentioned that the pairing energy E_{pair} is a measure for the size of pairing correlations in theoretical calculations, and under the zero-range pairing force it is calculated as

$$E_{\text{pair}} = -\frac{1}{2}\int d^3\boldsymbol{r}\kappa(\boldsymbol{r})\Delta(\boldsymbol{r}). \tag{15}$$

The quadrupole deformation is calculated using

$$\beta_2 = \frac{\sqrt{5\pi}Q_2}{3N\langle r^2\rangle} = \frac{2\sqrt{5\pi}\int d^3\boldsymbol{r}\rho_v(\boldsymbol{r})r^2P_2(\cos\theta)}{3[\int d^3\boldsymbol{r}\rho_v(\boldsymbol{r})][\int d^3\boldsymbol{r}\rho_v(\boldsymbol{r})r^2]}, \tag{16}$$

where N is the corresponding particle number, $\langle r^2\rangle$ is the root-mean-square (rms) radius, Q_2 is the intrinsic quadrupole moment, and ρ_v is the vector density distribution.

3. Numerical Details

In this work, the relativistic density functional PC-PK1 [71] is adopted, which leads to the root-mean-square deviation of 1.14 MeV in reproducing available mass data [75,76], among one of highest-accuracy density-functional descriptions for mass. The accuracy for mass descriptions by PC-PK1 for medium-heavy to superheavy regions reach several hundred keV, as illustrated in Refs. [57,77]. In the coordinate space, the box size $R_{\text{box}} = 20$ fm and the mesh size $\Delta r = 0.1$ fm. The cutoff parameters for the Dirac Woods-Saxon basis include the angular momentum cutoff $J_{\text{max}} = 23/2\,\hbar$ and the energy cutoff for positive-energy states $E^{+}_{\text{cut}} = 300$ MeV, and the number of negative-energy states is taken the same as that of positive-energy states [74]. For the Legendre expansion in Equation (12), a cutoff $\lambda_{\text{max}} = 10$ is taken. In the pairing channel, the pairing strength for Equation (10), $V_0 = -325$ MeV fm^3, and a pairing window of 100 MeV are taken. These pairing parameters are determined by reproducing the odd-even mass differences for Ca and Pb isotopes [64]. The examinations for the above numerical cutoffs and the pairing parameters can be found in Refs. [64,66].

4. Results and Discussions

The even-even nuclei in the $_{126}$Ubh isotopic chain are calculated with the DRHBc theory. Bulk properties for the ground states, such as binding energy, separation energies, quadrupole deformation, root-mean-square radii and pairing energies, as well as the corresponding density distributions and single-particle levels, are obtained. In order to discuss the possible magicity, in the following we will focus on the evolution of two-proton separation energy and two-proton gap, deformation and pairing energies, as well as single-particle levels.

The separation energy belongs to the first rang observables for magic numbers, and corresponds to the characteristic jumps in separation energies [78]. As has been shown in Figures 2 and 3 of Ref. [66], in the DRHBc calculations for $8 \leqslant Z \leqslant 120$, abrupt changes of separation energies are exhibited at magic numbers. In order to explore the possible magicity for superheavy nuclei, Figure 1a shows the two-proton separation energy S_{2p} in the DRHBc calculations by taking the even-even $N = 258$ isotopes with $Z > 100$ as examples. It is shown that S_{2p} decreases monotonously with the proton number, with the slopes larger near $Z = 120$ and smaller near $Z = 130$. For a more intuitive view, Figure 1b shows the corresponding decrease of S_{2p} at each Z, i.e., the two-proton gap, $\delta_{2p}(Z,N) = S_{2p}(Z,N) - S_{2p}(Z+2,N)$. The drastic changes of the two-nucleon separation energies are implicated by the peaks of the two-nucleon gaps, as has been shown in Figures 4 and 5 of Ref. [66]. Here a peak at $Z = 120$ is noted, which is consistent with several other density functional predictions that $Z = 120$ is a proton magic number. On the contrary, the δ_{2p} of $Z = 126$ corresponds to a decrease, instead of a peak, which has also been depicted in the slope of S_{2p}. Therefore, from the two-proton separation energy, $Z = 126$ does not possess extra stability, and no characteristic of magicity is shown.

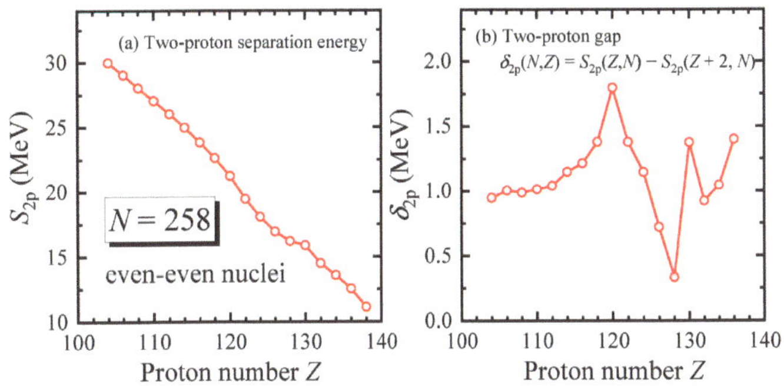

Figure 1. (**a**) Two-proton separation energy S_{2p} and (**b**) two-proton gap δ_{2p} as functions of the proton number in the DRHBc calculations with PC-PK1 for even-even $N = 258$ isotopes with $Z > 100$.

When the nucleon number equals to a traditional magic number, one result of RHB calculation is that at the lowest minimum of the potential energy curve, i.e., the ground state, usually has a spherical shape [66]. In order to further discuss the possibility of the proton magicity at $Z = 126$, Figure 2 shows the quadrupole deformation β_2 as a function of the neutron number in the DRHBc calculations for even-even $_{126}$Ubh isotopes. From the proton drip line and by increasing the neutron number, the deformation β_2 first generally decreases from a large positive value to zero near $N = 258$. After that, β_2 increases to a large value again and decreases to zero near $N = 350$. The β_2 at $324 \leqslant N \leqslant 334$ are missing because the corresponding isotopes are predicted to be unbound. According to the evolution of two-neutron separation energy and neutron Fermi energy, the isotopes at $336 \leqslant N \leqslant 350$ are bound again, which shows a similar behavior with the "re-entrant binding" in Refs. [2,57,58,79], and forms a "peninsula" beyond the primary two-neutron drip line at $N = 322$.

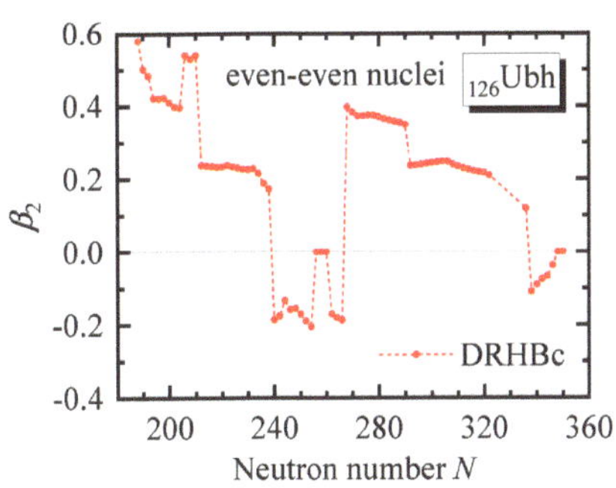

Figure 2. Quadrupole deformation β_2 as a function of the neutron number in the DRHBc calculations with PC-PK1 for even-even $_{126}$Ubh isotopes.

From the evolution of deformation, the information on nuclear structure can be extracted. Before focusing on the possible proton magicity, we first discuss the neutron structure, which is more obvious in the evolution with the increase of neutron number and can serve as a comparison for our discussion on the proton structure in the next paragraph. It is noted that except for a few spherical ones near $N = 258$ and 350, all remaining Ubh isotopes are deformed. The spherical Ubh isotopes near $N = 258$ and 350 correspond to new possible superheavy magic numbers. Sudden changes of β_2 occur near $N = 208, 240, 258, 268, 292$ and 338, which are also related to the evolution of shell structure and may lead to shape coexistence [60,61]. Taking $N = 266$ to 268 as an example, in Figure 2 a sudden change of deformation occurs between them. Figure 3 shows their potential energy curves from deformation-constrained DRHBc calculations by the solid curves, and the black squares represent the corresponding deformation minima from unconstrained DRHBc calculations. Both of them have two major minima at $\beta_2 \approx -0.2$ and 0.4, and a minor minimum occurs at $\beta_2 \approx 0.06$. These two major minima correspond to similar energies, supporting the possible shape coexistence. At $N = 266$ the oblate minimum is lower, and at $N = 268$ prolate minimum is lower, leading to the sudden change of β_2 at $N = 268$. In addition, considering that most Ubh isotopes are prolate rather than oblate, the prolate-shape dominance still works in the superheavy region [63]. Therefore, the evolution of neutron structure for Ubh isotopes is reflected on the quadrupole deformation, and the sphericity near $N = 258$ and 350 are significant signals corresponding to possible neutron magicity [80]. It should be also mentioned that Figure 2 focuses on the axial quadrupole deformation. When one wishes to study general stability properties in nuclear physics, especially of heavy nuclei, more quantities and effects, such as the deformation parameters with orders higher than quadrupole one, as well as the degrees of freedom from spatial-reflection asymmetry and triaxial shapes, also play important roles and should be taken into account [19,20,28,30].

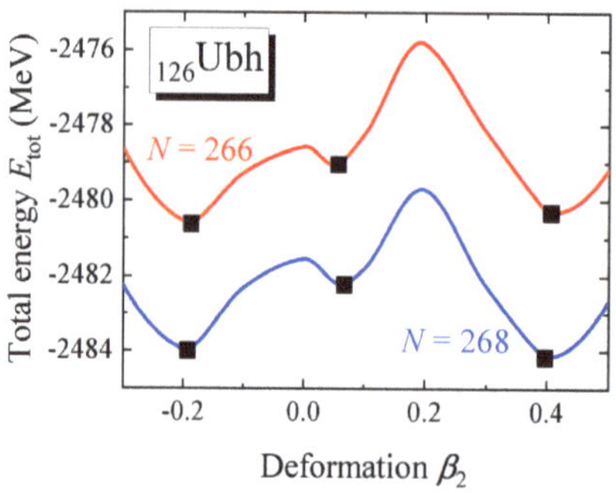

Figure 3. Potential energy curves at $N = 266$ and 268 in $_{126}$Ubh isotopic chain from the constrained DRHBc calculations. The black squares represent the corresponding deformation minima from unconstrained DRHBc calculations.

Then we come to the proton structure. As has been discussed above, a naïve analysis suggests $Z = 126$ to be a possible traditional proton magic number. The deformation shown in Figure 2 is significantly different with the behavior in the proton-magic isotopes with $Z \leqslant 120$. As shown in Ref. [66], in the isotopic chains with proton magic numbers $Z = 8, 20, 28, 50$, almost all nuclei are spherical, and in the isotopic chain with $Z = 82$, about half of the isotopes are spherical. In comparison, for $Z = 126$ here, only five isotopes near predicted neutron magic numbers $N = 258$ and 350 are spherical. In conclusion, not only the evolutions of S_{2p} and δ_{2p} depicted in Figure 1, but also the behavior of β_2 in Figure 2, do not support the proton magic number $Z = 126$. It should also be noted that the conclusion drawn here is limited by the choice of one single axial deformation parameter β_2. Higher orders such as β_4 and β_6, as well as the reflection asymmetry degrees of freedom such like β_3 and β_5 which has not been incorporated in the DRHBc theory yet, also impact the calculations [81,82], and are expected to be studied in future works.

For magic nuclei, a notable result from RHB calculations is that a large gap exists in the single-particle spectrum, which further implies that the occupied nucleons cannot be scattered into higher orbits, resulting in the vanishing of pairing energy. Conversely, the sudden disappearance of pairing energy may serve as a theoretical signal to aid in the search for possible magic numbers. Figure 4 shows the neutron and proton pairing energies as functions of the neutron number in the DRHBc calculations for even-even $_{126}$Ubh isotopes.

The evolution of neutron pairing energy is smooth at $268 \leqslant N \leqslant 320$, except a few turning points, while at other neutron numbers, a significant staggering pattern occurs between zero and nonzero values, such as the results near $N = 240$ and 340. Since we only consider even-even Ubh nuclei in this work, the staggerings in neutron pairing energy are irrelevant to odd-even effect. The zero values at $N = 258$ and 350 correspond to the predicted neutron magic numbers, while those at $N = 232, 240, 248, 320, 336, 344$ reflect the evolution of neutron shell structure in deformed nuclei.

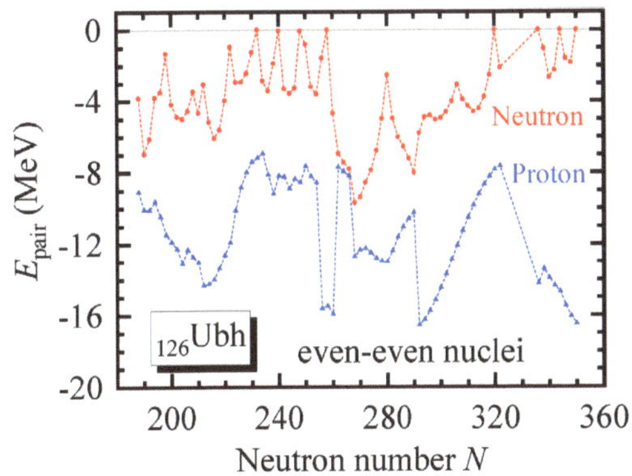

Figure 4. Neutron and proton pairing energies as functions of the neutron number in the DRHBc calculations with PC-PK1 for even-even $_{126}$Ubh isotopes.

The evolution of proton pairing energy is generally smooth for most Ubh isotopes, except for a few turning points. Its behavior shows no notable staggering, and never vanishes, which is remarkably different from the neutron one. For the spherical nuclei near $N = 258$ and 350, the corresponding proton pairing energies are significantly larger than their neighbors. Therefore, the analysis on pairing energies does not support the proton magic number $Z = 126$.

In order to further discuss the possibility of proton magicity at $Z = 126$, taking the spherical isotope $^{384}_{126}$Ubh as an example, Figure 5 shows the single proton levels around the Fermi energy λ_p in the canonical basis. If we neglect the pairing correlation and let the proton occupies the orbits one by one, $Z = 126$ would fully occupy the $i_{11/2}$ orbit at $\epsilon_p = -7.847$. This orbit is very close to the $p_{1/2}$ above and the $p_{3/2}$ below, resulting in no discernible gap formed in the single-proton spectrum. Although the neutron magic number $N = 126$ has been confirmed and successfully reproduced by the DRHBc calculations [58], $Z = 126$ is not supported to be a magic number here.

It should also be noticed that two significant gaps appear at $Z = 120$ and 138, consistent with the prediction of proton magic number $Z = 120$ from the evolution of S_{2p} and δ_{2p} in Figure 1. As has been discussed above, the DRHBc calculations have self-consistently reproduced all traditional magic numbers, and the prediction on $Z = 120$ and 138 is also consistent with existing literature, such as Refs. [44,83]. We also expect further DRHBc studies on other superheavy isotopic chains to confirm the prediction of these new magic numbers.

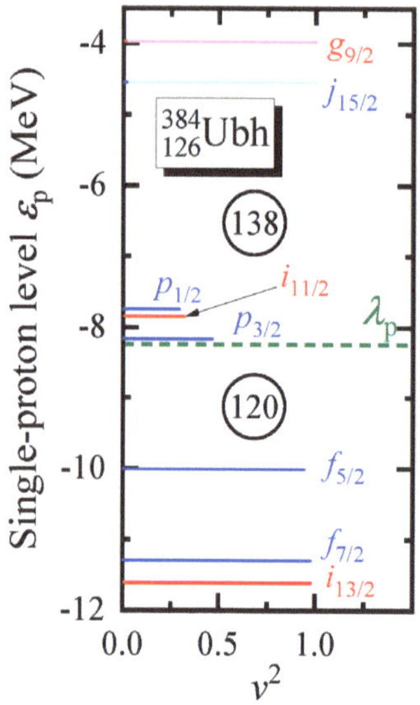

Figure 5. Single-proton levels around the Fermi energy λ_p in the canonical basis for $^{384}_{126}$Ubh versus the occupation probability v^2 in the DRHBc calculations with PC-PK1. Each level is labeled by the quantum numbers for spherical orbits. The Fermi energy λ_p is shown as a green dotted line.

5. Summary

The even-even $_{126}$Ubh isotopes are calculated based on the DRHBc calculations with PC-PK1. The evolutions of quadrupole deformation and pairing energies for neutron and proton are analyzed to study the possible nuclear magicity. Spherical shape occurs and neutron pairing energy vanishes at $N = 258$ and 350, supporting that they are neutron magic numbers, while the proton pairing energy never vanishes in Ubh isotopes. In the single-proton spectrum, there is no discernible gap at $Z = 126$, while significant gaps appear at $Z = 120$ and 138. Therefore, $Z = 126$ is not supported as a proton magic number, while $Z = 120$ and 138 are suggested as candidates of proton magic numbers.

We expect further DRHBc studies on other superheavy isotopic chains to provide the corresponding ground-state properties, so that we can extract the two-proton gaps, which can be used as one of the signatures for magic numbers [44,83], to confirm the conclusions on proton magic numbers in this work. In addition, in the present studies we have only considered the ground-state static properties related to energies and shapes, while the α, β decays or fissions are not discussed yet. It should also be noted that the conclusion drawn here is limited by the choice of one single axial deformation parameter β_2, while higher orders such as β_4 and β_6 are not discussed in this work, and the reflection asymmetry degrees of freedom such like β_3 and β_5 is not incorporated in the present DRHBc theory. They may significantly influence the stability in superheavy nuclei, and their effects are expected to be investigated in future works.

Author Contributions: Conceptualization, C.P.; methodology, C.P. and X.-H.W.; formal analysis, C.P. and X.-H.W.; investigation, C.P. and X.-H.W.; writing—original draft preparation, C.P.; writing—review and editing, All authors; visualization, C.P.; project administration, C.P.; funding acquisition, X.-H.W.; All authors have read and agreed to the published version of the manuscript.

Funding: This work was partly supported by the National Natural Science Foundation of China under Grant No. 12405134 and the start-up grant XRC-23103 of Fuzhou University.

Data Availability Statement: The dataset can be accessed upon request to the corresponding author.

Acknowledgments: Helpful discussions with members of the DRHBc Mass Table Collaboration are highly appreciated. C.P. would like to express gratitude to Y. Kim and M. Mun for the generous help during the stay at CENS, IBS.

Conflicts of Interest: The authors declare no conflicts of interest.

References

1. Thoennessen, M. Current status and future potential of nuclide discoveries. *Rep. Prog. Phys.* **2013**, *76*, 056301. [CrossRef]
2. Erler, J.; Birge, N.; Kortelainen, M.; Nazarewicz, W.; Olsen, E.; Perhac, A.M.; Stoitsov, M. The limits of the nuclear landscape. *Nature* **2012**, *486*, 509–512. [CrossRef] [PubMed]
3. Moller, P. The limits of the nuclear chart set by fission andalpha decay. In Proceedings of the Nobel Symposium NS160—Chemistry and Physics of Heavy and Superheavy Elements, Kristianstad, Sweden, 29 May–3 June 2016; Volume 131, p. 03002. [CrossRef]
4. Meng, J. (Ed.) *Relativistic Density Functional for Nuclear Structure*; World Scientific: Singapore, 2016. [CrossRef]
5. Zhang, Z.Y.; Gan, Z.G.; Yang, H.B.; Ma, L.; Huang, M.H.; Yang, C.L.; Zhang, M.M.; Tian, Y.L.; Wang, Y.S.; Sun, M.D.; et al. New Isotope ^{220}Np: Probing the Robustness of the $N = 126$ Shell Closure in Neptunium. *Phys. Rev. Lett.* **2019**, *122*, 192503. [CrossRef] [PubMed]
6. Ahn, D.S.; Fukuda, N.; Geissel, H.; Inabe, N.; Iwasa, N.; Kubo, T.; Kusaka, K.; Morrissey, D.J.; Murai, D.; Nakamura, T.; et al. Location of the Neutron Dripline at Fluorine and Neon. *Phys. Rev. Lett.* **2019**, *123*, 212501. [CrossRef]
7. Tanihata, I.; Hamagaki, H.; Hashimoto, O.; Shida, Y.; Yoshikawa, N.; Sugimoto, K.; Yamakawa, O.; Kobayashi, T.; Takahashi, N. Measurements of Interaction Cross Sections and Nuclear Radii in the Light p-Shell Region. *Phys. Rev. Lett.* **1985**, *55*, 2676–2679. [CrossRef] [PubMed]
8. Ozawa, A.; Kobayashi, T.; Suzuki, T.; Yoshida, K.; Tanihata, I. New Magic Number, $N = 16$, near the Neutron Drip Line. *Phys. Rev. Lett.* **2000**, *84*, 5493–5495. [CrossRef]
9. Adrich, P.; Klimkiewicz, A.; Fallot, M.; Boretzky, K.; Aumann, T.; Cortina-Gil, D.; Pramanik, U.D.; Elze, T.W.; Emling, H.; Geissel, H.; et al. Evidence for Pygmy and Giant Dipole Resonances in ^{130}Sn and ^{132}Sn. *Phys. Rev. Lett.* **2005**, *95*, 132501. [CrossRef] [PubMed]
10. Oganessian, Y.T.; Abdullin, F.S.; Alexander, C.; Binder, J.; Boll, R.A.; Dmitriev, S.N.; Ezold, J.; Felker, K.; Gostic, J.M.; Grzywacz, R.K.; et al. Production and Decay of the Heaviest Nuclei ^{293,294}117 and ^{294}118. *Phys. Rev. Lett.* **2012**, *109*, 162501. [CrossRef] [PubMed]
11. Sobiczewski, A.; Gareev, F.A.; Kalinkin, B.N. Closed shells for $Z > 82$ and $N > 126$ in a diffuse potential well. *Phys. Lett.* **1966**, *22*, 500–502. [CrossRef]
12. Meldner, H. Predictions of new magic regions and masses for super heavy nuclei from calculations with realistic shell model single particle Hamiltonians. In Proceedings of the Lysekil Symposium: Nuclides far off the Stability Line, Lysekil, Sweden, 21–27 August 1967; Volume 36, p. 593.
13. Nilsson, S.G.; Tsang, C.F.; Sobiczewski, A.; Szymański, Z.; Wycech, S.; Gustafson, C.; Lamm, I.L.; Möller, P.; Nilsson, B. On the nuclear structure and stability of heavy and superheavy elements. *Nucl. Phys. A* **1969**, *131*, 1–66. [CrossRef]
14. Mosel, U.; Greiner, W. On the stability of superheavy nuclei against fission. *Zeit. Phys. A* **1969**, *222*, 261. [CrossRef]
15. National Nuclear Data Center (NNDC). Available online: https://www.nndc.bnl.gov/ (accessed on 25 December 2024).
16. Mayer, M.G. On Closed Shells in Nuclei. II. *Phys. Rev.* **1949**, *75*, 1969–1970. [CrossRef]
17. Haxel, O.; Jensen, J.H.D.; Suess, H.E. On the "Magic Numbers" in Nuclear Structure. *Phys. Rev.* **1949**, *75*, 1766–1766. [CrossRef]
18. Yang, J.; Dudek, J.; Dedes, I.; Baran, A.; Curien, D.; Gaamouci, A.; Góźdź, A.; Pędrak, A.; Rouvel, D.; Wang, H.L.; et al. Exotic shape symmetries around the fourfold octupole magic number $N = 136$: Formulation of experimental identification criteria. *Phys. Rev. C* **2022**, *105*, 034348. [CrossRef]
19. Yang, J.; Dudek, J.; Dedes, I.; Baran, A.; Curien, D.; Gaamouci, A.; Góźdź, A.; Pędrak, A.; Rouvel, D.; Wang, H.L. Exotic symmetries as stabilizing factors for superheavy nuclei: Symmetry-oriented generalized concept of nuclear magic numbers. *Phys. Rev. C* **2022**, *106*, 054314. [CrossRef]
20. Yang, J.; Dudek, J.; Dedes, I.; Baran, A.; Curien, D.; Gaamouci, A.; Góźdź, A.; Pędrak, A.; Rouvel, D.; Wang, H.L. Islands of oblate hyperdeformed and superdeformed superheavy nuclei with D_{3h} point group symmetry in competition with normal-deformed D_{3h} states: "Archipelago" of D_{3h}-symmetry islands. *Phys. Rev. C* **2023**, *107*, 054304. [CrossRef]
21. Wheeler, J.A. Nuclear fission and nuclear stability. In *Niels Bohr and the Development of Physics: Essays Dedicated to Niels Bohr on the Occasion of His Seventieth Birthday*; Pauli, W.; Rosenfeld, L.; Weisskopf, V., Eds.; McGraw-Hill: New York, NY, USA, 1955; p. 163.
22. Werner, F.G.; Wheeler, J.A. Superheavy Nuclei. *Phys. Rev.* **1958**, *109*, 126–144. [CrossRef]
23. Myers, W.D.; Swiatecki, W.J. Nuclear masses and deformations. *Nucl. Phys.* **1966**, *81*, 1–60. [CrossRef]
24. Siemens, P.J.; Bethe, H.A. Shape of Heavy Nuclei. *Phys. Rev. Lett.* **1967**, *18*, 704–706. [CrossRef]

25. Ćwiok, S.; Dobaczewski, J.; Heenen, P.H.; Magierski, P.; Nazarewicz, W. Shell structure of the superheavy elements. *Nucl. Phys. A* **1996**, *611*, 211–246. [CrossRef]
26. Moller, P.; Nix, J.R. Stability of heavy and superheavy elements. *J. Phys. G* **1994**, *20*, 1681. [CrossRef]
27. Dedes, I.; Dudek, J. Propagation of the nuclear mean-field uncertainties with increasing distance from the parameter adjustment zone: Applications to superheavy nuclei. *Phys. Rev. C* **2019**, *99*, 054310. [CrossRef]
28. Gaamouci, A.; Dedes, I.; Dudek, J.; Baran, A.; Benhamouda, N.; Curien, D.; Wang, H.L.; Yang, J. Exotic toroidal and superdeformed configurations in light atomic nuclei: Predictions using a mean-field Hamiltonian without parametric correlations. *Phys. Rev. C* **2021**, *103*, 054311. [CrossRef]
29. Agbemava, S.E.; Afanasjev, A.V.; Ray, D.; Ring, P. Assessing theoretical uncertainties in fission barriers of superheavy nuclei. *Phys. Rev. C* **2017**, *95*, 054324. [CrossRef]
30. Agbemava, S.E.; Afanasjev, A.V.; Taninah, A. Propagation of statistical uncertainties in covariant density functional theory: Ground state observables and single-particle properties. *Phys. Rev. C* **2019**, *99*, 014318. [CrossRef]
31. Cao, L.G.; Ma, Z.Y. Exploration of resonant continuum and giant resonance in the relativistic approach. *Phys. Rev. C* **2002**, *66*, 024311. [CrossRef]
32. Meng, J.; Toki, H.; Zhou, S.G.; Zhang, S.Q.; Long, W.H.; Geng, L.S. Relativistic continuum Hartree Bogoliubov theory for ground-state properties of exotic nuclei. *Prog. Part. Nucl. Phys.* **2006**, *57*, 470–563. [CrossRef]
33. Dobaczewski, J.; Michel, N.; Nazarewicz, W.; Płoszajczak, M.; Rotureau, J. Shell structure of exotic nuclei. *Prog. Part. Nucl. Phys.* **2007**, *59*, 432–445. [CrossRef]
34. Pei, J.C.; Kruppa, A.T.; Nazarewicz, W. Quasiparticle continuum and resonances in the Hartree-Fock-Bogoliubov theory. *Phys. Rev. C* **2011**, *84*, 024311. [CrossRef]
35. Dobaczewski, J.; Flocard, H.; Treiner, J. Hartree-Fock-Bogolyubov description of nuclei near the neutron-drip line. *Nucl. Phys. A* **1984**, *422*, 103–139. [CrossRef]
36. Terasaki, J.; Heenen, P.H.; Flocard, H.; Bonche, P. 3D solution of Hartree-Fock-Bogoliubov equations for drip-line nuclei. *Nucl. Phys. A* **1996**, *600*, 371–386. [CrossRef]
37. Pöschl, W.; Vretenar, D.; Lalazissis, G.A.; Ring, P. Relativistic Hartree-Bogoliubov Theory with Finite Range Pairing Forces in Coordinate Space: Neutron Halo in Light Nuclei. *Phys. Rev. Lett.* **1997**, *79*, 3841–3844. [CrossRef]
38. Meng, J. Relativistic continuum Hartree-Bogoliubov theory with both zero range and finite range Gogny force and their application. *Nucl. Phys. A* **1998**, *635*, 3–42. [CrossRef]
39. Meng, J.; Ring, P. Relativistic Hartree-Bogoliubov Description of the Neutron Halo in ^{11}Li. *Phys. Rev. Lett.* **1996**, *77*, 3963–3966. [CrossRef]
40. Meng, J.; Ring, P. Giant Halo at the Neutron Drip Line. *Phys. Rev. Lett.* **1998**, *80*, 460–463. [CrossRef]
41. Meng, J.; Sugawara-Tanabe, K.; Yamaji, S.; Ring, P.; Arima, A. Pseudospin symmetry in relativistic mean field theory. *Phys. Rev. C* **1998**, *58*, R628–R631. [CrossRef]
42. Meng, J.; Tanihata, I.; Yamaji, S. The proton and neutron distributions in Na isotopes: the development of halo and shell structure. *Phys. Lett. B* **1998**, *419*, 1–6. [CrossRef]
43. Meng, J.; Zhou, S.G.; Tanihata, I. The relativistic continuum Hartree-Bogoliubov description of charge-changing cross section for C, N, O and F isotopes. *Phys. Lett. B* **2002**, *532*, 209–214. [CrossRef]
44. Zhang, W.; Meng, J.; Zhang, S.Q.; Geng, L.S.; Toki, H. Magic numbers for superheavy nuclei in relativistic continuum Hartree-Bogoliubov theory. *Nucl. Phys. A* **2005**, *753*, 106–135. [CrossRef]
45. Xia, X.W.; Lim, Y.; Zhao, P.W.; Liang, H.Z.; Qu, X.Y.; Chen, Y.; Liu, H.; Zhang, L.F.; Zhang, S.Q.; Kim, Y.; et al. The limits of the nuclear landscape explored by the relativistic continuum Hartree-Bogoliubov theory. *At. Data Nucl. Data Tables* **2018**, *121-122*, 1–215. [CrossRef]
46. Zhou, S.G.; Meng, J.; Ring, P.; Zhao, E.G. Neutron halo in deformed nuclei. *Phys. Rev. C* **2010**, *82*, 011301. [CrossRef]
47. Li, L.; Meng, J.; Ring, P.; Zhao, E.G.; Zhou, S.G. Deformed relativistic Hartree-Bogoliubov theory in continuum. *Phys. Rev. C* **2012**, *85*, 024312. [CrossRef]
48. Sun, X.X.; Zhao, J.; Zhou, S.G. Shrunk halo and quenched shell gap at $N = 16$ in ^{22}C: Inversion of sd states and deformation effects. *Phys. Lett. B* **2018**, *785*, 530–535. [CrossRef]
49. Sun, X.X.; Zhao, J.; Zhou, S.G. Study of ground state properties of carbon isotopes with deformed relativistic Hartree-Bogoliubov theory in continuum. *Nucl. Phys. A* **2020**, *1003*, 122011. [CrossRef]
50. Yang, Z.H.; Kubota, Y.; Corsi, A.; Yoshida, K.; Sun, X.X.; Li, J.G.; Kimura, M.; Michel, N.; Ogata, K.; Yuan, C.X.; et al. Quasifree Neutron Knockout Reaction Reveals a Small *s*-Orbital Component in the Borromean Nucleus ^{17}B. *Phys. Rev. Lett.* **2021**, *126*, 082501. [CrossRef] [PubMed]
51. Sun, X.X. Deformed two-neutron halo in ^{19}B. *Phys. Rev. C* **2021**, *103*, 054315. [CrossRef]
52. Zhong, S.Y.; Zhang, S.S.; Sun, X.X.; Smith, M.S. Study of the deformed halo nucleus ^{31}Ne with Glauber model based on microscopic self-consistent structures. *Sci. China Phys. Mech. Astron.* **2022**, *65*, 262011. [CrossRef]

53. Zhang, K.Y.; Papakonstantinou, P.; Mun, M.H.; Kim, Y.; Yan, H.; Sun, X.X. Collapse of the $N = 28$ shell closure in the newly discovered ^{39}Na nucleus and the development of deformed halos towards the neutron dripline. *Phys. Rev. C* **2023**, *107*, L041303. [CrossRef]
54. Zhang, K.Y.; Yang, S.Q.; An, J.L.; Zhang, S.S.; Papakonstantinou, P.; Mun, M.H.; Kim, Y.; Yan, H. Missed prediction of the neutron halo in ^{37}Mg. *Phys. Lett. B* **2023**, *844*, 138112. [CrossRef]
55. An, J.L.; Zhang, K.Y.; Lu, Q.; Zhong, S.Y.; Zhang, S.S. A unified description of the halo nucleus ^{37}Mg from microscopic structure to reaction observables. *Phys. Lett. B* **2024**, *849*, 138422. [CrossRef]
56. Pan, C.; Zhang, K.; Zhang, S. Nuclear magnetism in the deformed halo nucleus ^{31}Ne. *Phys. Lett. B* **2024**, *855*, 138792. [CrossRef]
57. Zhang, K.; He, X.; Meng, J.; Pan, C.; Shen, C.; Wang, C.; Zhang, S. Predictive power for superheavy nuclear mass and possible stability beyond the neutron drip line in deformed relativistic Hartree-Bogoliubov theory in continuum. *Phys. Rev. C* **2021**, *104*, L021301. [CrossRef]
58. Pan, C.; Zhang, K.Y.; Chong, P.S.; Heo, C.; Ho, M.C.; Lee, J.; Li, Z.P.; Sun, W.; Tam, C.K.; Wong, S.H.; et al. Possible bound nuclei beyond the two-neutron drip line in the $50 \leq Z \leq 70$ region. *Phys. Rev. C* **2021**, *104*, 024331. [CrossRef]
59. He, X.T.; Wu, J.W.; Zhang, K.Y.; Shen, C.W. Odd-even differences in the stability "peninsula" in the $106 \leq Z \leq 112$ region with the deformed relativistic Hartree-Bogoliubov theory in continuum. *Phys. Rev. C* **2024**, *110*, 014301. [CrossRef]
60. Choi, Y.B.; Lee, C.H.; Mun, M.H.; Kim, Y. Bubble nuclei with shape coexistence in even-even isotopes of Hf to Hg. *Phys. Rev. C* **2022**, *105*, 024306. [CrossRef]
61. Kim, S.; Mun, M.H.; Cheoun, M.K.; Ha, E. Shape coexistence and neutron skin thickness of Pb isotopes by the deformed relativistic Hartree-Bogoliubov theory in continuum. *Phys. Rev. C* **2022**, *105*, 034340. [CrossRef]
62. Mun, M.H.; Kim, S.; Cheoun, M.K.; So, W.Y.; Choi, S.; Ha, E. Odd-even shape staggering and kink structure of charge radii of Hg isotopes by the deformed relativistic Hartree-Bogoliubov theory in continuum. *Phys. Lett. B* **2023**, *847*, 138298. [CrossRef]
63. Guo, P.; Pan, C.; Zhao, Y.C.; Du, X.K.; Zhang, S.Q. Prolate-shape dominance in atomic nuclei within the deformed relativistic Hartree-Bogoliubov theory in continuum. *Phys. Rev. C* **2023**, *108*, 014319. [CrossRef]
64. Zhang, K.; Cheoun, M.K.; Choi, Y.B.; Chong, P.S.; Dong, J.; Geng, L.; Ha, E.; He, X.; Heo, C.; Ho, M.C.; et al. Deformed relativistic Hartree-Bogoliubov theory in continuum with a point-coupling functional: Examples of even-even Nd isotopes. *Phys. Rev. C* **2020**, *102*, 024314. [CrossRef]
65. Pan, C.; Cheoun, M.K.; Choi, Y.B.; Dong, J.; Du, X.; Fan, X.H.; Gao, W.; Geng, L.; Ha, E.; He, X.T.; et al. Deformed relativistic Hartree-Bogoliubov theory in continuum with a point-coupling functional. II. Examples of odd Nd isotopes. *Phys. Rev. C* **2022**, *106*, 014316. [CrossRef]
66. Zhang, K.; Cheoun, M.K.; Choi, Y.B.; Chong, P.S.; Dong, J.; Dong, Z.; Du, X.; Geng, L.; Ha, E.; He, X.T.; et al. Nuclear mass table in deformed relativistic Hartree-Bogoliubov theory in continuum, I: Even-even nuclei. *At. Data Nucl. Data Tables* **2022**, *144*, 101488. [CrossRef]
67. Guo, P.; Cao, X.; Chen, K.; Chen, Z.; Cheoun, M.K.; Choi, Y.B.; Lam, P.C.; Deng, W.; Dong, J.; Du, P.; et al. Nuclear mass table in deformed relativistic Hartree-Bogoliubov theory in continuum, II: Even-Z nuclei. *At. Data Nucl. Data Tables* **2024**, *158*, 101661. [CrossRef]
68. Wang, M.; Huang, W.J.; Kondev, F.G.; Audi, G.; Naimi, S. The AME 2020 atomic mass evaluation (II). Tables, graphs and references. *Chin. Phys. C* **2021**, *45*, 030003. [CrossRef]
69. Chatt, J. Recommendations for the Naming of Elements of Atomic Numbers Greater than 100. *Pure Appl. Chem.* **1979**, *51*, 381–384. [CrossRef]
70. Bürvenich, T.; Madland, D.G.; Maruhn, J.A.; Reinhard, P.G. Nuclear ground state observables and QCD scaling in a refined relativistic point coupling model. *Phys. Rev. C* **2002**, *65*, 044308. [CrossRef]
71. Zhao, P.W.; Li, Z.P.; Yao, J.M.; Meng, J. New parametrization for the nuclear covariant energy density functional with a point-coupling interaction. *Phys. Rev. C* **2010**, *82*, 054319. [CrossRef]
72. Kucharek, H.; Ring, P. Relativistic field theory of superfluidity in nuclei. *Z. Phys. A* **1991**, *339*, 23–35. [CrossRef]
73. Ring, P.; Schuck, P. *The Nuclear Many-Body Problem*; Springer: Berlin/Heidelberg, Germany, 1980.
74. Zhou, S.G.; Meng, J.; Ring, P. Spherical relativistic Hartree theory in a Woods-Saxon basis. *Phys. Rev. C* **2003**, *68*, 034323. [CrossRef]
75. Zhang, Q.S.; Niu, Z.M.; Li, Z.P.; Yao, J.M.; Meng, J. Global dynamical correlation energies in covariant density functional theory: Cranking approximation. *Front. Phys.* **2014**, *9*, 529–536. [CrossRef]
76. Lu, K.Q.; Li, Z.X.; Li, Z.P.; Yao, J.M.; Meng, J. Global study of beyond-mean-field correlation energies in covariant energy density functional theory using a collective Hamiltonian method. *Phys. Rev. C* **2015**, *91*, 027304. [CrossRef]
77. Zhao, P.W.; Song, L.S.; Sun, B.; Geissel, H.; Meng, J. Crucial test for covariant density functional theory with new and accurate mass measurements from Sn to Pa. *Phys. Rev. C* **2012**, *86*, 064324. [CrossRef]
78. Dudek, J. Reexamination of the Nuclear Stability of $Z = 114$ to $Z = 126$ Superheavy Nuclei with the Use of the Deformed Woods-Saxon Potential. *Acta Phys. Pol. B* **1978**, *9*, 919.

79. Stoitsov, M.V.; Dobaczewski, J.; Nazarewicz, W.; Pittel, S.; Dean, D.J. Systematic study of deformed nuclei at the drip lines and beyond. *Phys. Rev. C* **2003**, *68*, 054312. [CrossRef]
80. Liu, W.J.; Lv, C.J.; Guo, P.; Pan, C.; Wang, S.; Wu, X.H. Magic Number $N = 350$ Predicted by the Deformed Relativistic Hartree-Bogoliubov Theory in Continuum: $Z = 136$ Isotopes as an Example. *Particles* **2024**, *7*, 1078–1085. [CrossRef]
81. Jachimowicz, P.; Kowal, M.; Skalski, J. Adiabatic fission barriers in superheavy nuclei. *Phys. Rev. C* **2017**, *95*, 014303. [CrossRef]
82. Rahmatinejad, A.; Bezbakh, A.N.; Shneidman, T.M.; Adamian, G.; Antonenko, N.V.; Jachimowicz, P.; Kowal, M. Level-density parameters in superheavy nuclei. *Phys. Rev. C* **2021**, *103*, 034309. [CrossRef]
83. Li, J.J.; Long, W.H.; Margueron, J.; Van Giai, N. Superheavy magic structures in the relativistic Hartree-Fock-Bogoliubov approach. *Phys. Lett. B* **2014**, *732*, 169–173. [CrossRef]

particles

MDPI

Article

Exploring the Neutron Magic Number in Superheavy Nuclei: Insights into $N = 258$

Pengxiang Du and Jian Li *

College of Physics, Jilin University, Changchun 130012, China; dupx24@mails.jlu.edu.cn
* Correspondence: jianli@jlu.edu.cn

Abstract: In the framework of axial symmetric relativistic Hartree–Bogoliubov (RHB) theory and the Skyrme Hartree–Fock–Bogoliubov (HFB) theory, the evolution of shell structure, density distribution, and ground state deformation in superheavy nuclei proximate to $N = 258$ are investigated within the relativistic functionals DD-PC1 and DD-ME2, as well as the non-relativistic functional UNEDF0. The results from DD-ME2 and UNEDF0 indicate that $N = 258$ is a neutron magic number, whereas DD-PC1 does not anticipate the existence of a bound $N = 258$ magic nucleus. Further discussion suggests that the emergence of the magic number $N = 258$ is related to the depression of the central density.

Keywords: superheavy nuclei; magic number; relativistic Hartree–Bogoliubov approach; Skyrme Hartree–Fock–Bogoliubov approach

1. Introduction

The synthesis of superheavy elements and the exploration of their stability are important research areas in low-energy nuclear physics, which will help us address fundamental questions such as the boundaries of the nuclear chart and the limits of element existence. The heaviest element observed so far has a proton number $Z = 118$ [1–3], and attempts to synthesize even heavier elements are ongoing [4–7]. For superheavy nuclei, their stability is primarily determined by shell effects, and self-consistent mean-field models based on various relativistic and non-relativistic energy density functionals [8–12] are powerful tools for studying the shell structure and stability of these nuclei.

Over the past few decades, numerous systematic studies of the superheavy nuclei region have been conducted using EDFs. Early calculations are based on several sets of relativistic and non-relativistic functionals under the assumption of spherical symmetry and have explored the distribution of magic numbers in superheavy nuclei [13–16]. The predictions of different functionals show some variation, such as proton numbers $Z = 114$ or $Z = 120$, and neutron numbers $N = 172$ and $N = 184$. However, most investigations predict the appearance of a spherical shell gap at $N = 258$. Ref. [17] reveals the influence of spin–orbit splitting on the formation of the shell gap at $Z = 120$ based on relativistic Hartree–Fock–Bogoliubov (RHFB) calculations. Ref. [18] discusses the impact of the evolution of central density in superheavy nuclei on the shell structure, with results showing that the depression of nuclear central density leads to shell gaps at $Z = 120$ and $N = 172$. In contrast, a flatter density distribution is more favorable for the appearance of the magic number $N = 184$. The RHB calculations under the assumption of axial deformation reveal the significant role of deformation in the shell evolution of superheavy nuclei [19]. The results indicate that the shell gap at $N = 172$ has a minimal impact on the structure of superheavy nuclei, whereas the shell gap at $N = 184$ has a substantial impact. Ref. [20], based on RHB calculations with axial and triaxial deformation, predicts the widespread existence of toroidal energy minima in superheavy nuclei and forecasts the existence of spherical regions of superheavy nuclei. Nevertheless, none of these studies discuss the potential existence of the larger neutron magic number $N = 258$.

Recently, Ref. [21] conducted a systematic study on the ground-state and fission properties of actinides and superheavy elements using various relativistic density functionals, revealing a shell gap at $N = 258$. Meanwhile, calculations using the PC-PK1 functional [22] within the framework of deformed relativistic Hartree–Bogoliubov in continuum theory [23] have supported 258 as a neutron magic number [24–26]. To investigate whether 258 is indeed a neutron magic number, this work explores the shell structure evolution near $N = 258$ within the axial RHB and HFB frameworks, based on the relativistic density functionals DD-PC1 [27] and DD-ME2 [28], as well as the non-relativistic density functional UNEDF0 [29].

The article is arranged as follows. In Section 2, the numerical conditions employed in the calculations using the RHB and Skyrme HFB methods are elaborated. The results for the two-neutron shell gaps near $N = 258$, the evolution of the spherical shell structure, and the density distributions obtained from calculations based on different density functionals are presented and discussed in Section 3. The Section 4 provides a summary and offers some perspectives.

2. The Details of the Theoretical Calculations

In the calculations performed within the axial RHB framework [30], a separable form of the finite-range Gogny pairing interaction [31] is adopted to avoid uncertainties arising from the choice of the pairing window [32]. Due to the omission of fission, the selection of ground state is restricted to a certain range of quadrupole deformations. Through multiple tests, it is found that truncating the basis to $N_F = 20$ fermionic shells and $N_B = 20$ bosonic shells already yields satisfactory accuracy.

The calculations based on the UNEDF0 functional are performed within the axial Skyrme HFB framework [33,34]. A density-dependent zero-range force with a mixture of volume and surface characteristics is chosen for the pairing interaction. To avoid pairing collapse near closed shells, the Lipkin-Nogami (LN) method is employed in the calculations [29,35,36]. After extensive testing, the quasiparticle energy cutoff is set to 60 MeV, and the basis is truncated to 26 shells.

3. Results and Discussion

The peak in the two-nucleon gaps represents the sharp change in the two-nucleon separation energy, which can be regarded as a signature of the emergence of magic numbers [16,17]. The two-neutron gap δ_{2n} is defined as

$$\delta_{2n}(Z, N) = S_{2n}(Z, N) - S_{2n}(Z, N+2), \tag{1}$$

where $S_{2n}(Z, N)$ is the two neutron separation energy for a nucleus with proton number Z and neutron number N. The two-neutron gap δ_{2n} essentially reflects the rate of change in the binding energy of nuclei along an isotopic chain.

The two-neutron gaps and ground-state deformations obtained from RHB calculations using the DD-PC1 and DD-ME2 functionals and HFB calculations using the UNEDF0 functional are shown in Figure 1. The hollow symbols represent unbound nuclei. The calculations based on the DD-PC1 functional cover the isotopic chains from $Z = 96$ to $Z = 110$. The results indicate that DD-PC1 does not yield a bound nucleus at $N = 258$ within the calculated isotopic chains. Specifically, for the $Z = 110$ chain, the drip line is located at $N = 256$.

For the DD-ME2 functional, the calculations span the isotopic chains from $Z = 112$ to $Z = 126$. In Figure 1b, a strong shell effect is also observed at $N = 258$, where the two-neutron shell gap gradually decreases with the increasing proton number. It significantly reduces at $Z = 124$, and the peak disappears entirely at $Z = 126$. Meanwhile, a prominent peak appears at $N = 264$. In Figure 1e, the evolution of the deformation at the ground state shows that the competition between spherical and prolate shapes leads to the disappearance of the shell structure at $Z = 126$. On the other hand, for the chains from $Z = 112$ to $Z = 120$, the nuclei at $N = 260$ are unbound. For $Z = 120$, the nucleus at $N = 264$ becomes bound, likely due to the emergence of a new shell structure at $N = 264$, which increases the stability of nearby nuclei.

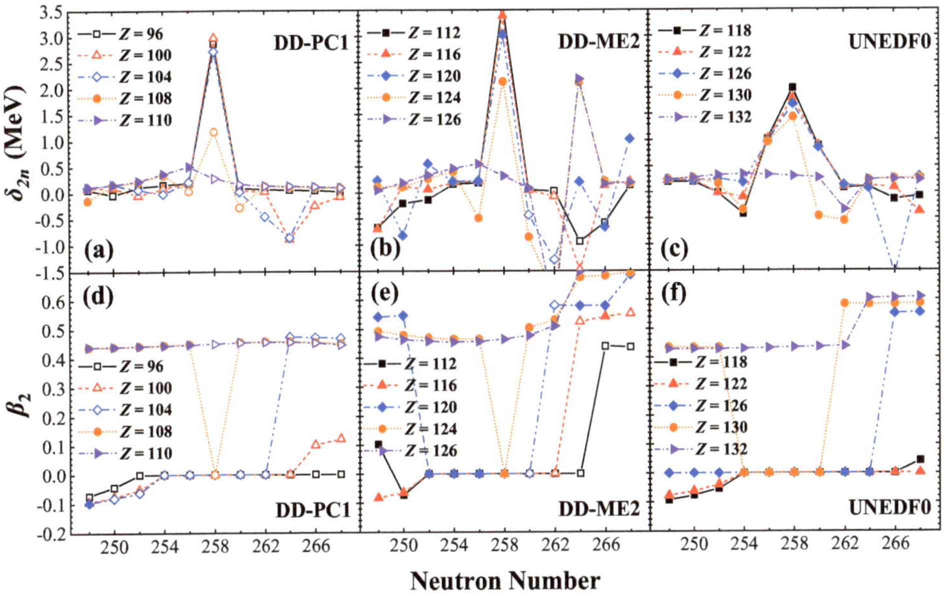

Figure 1. The two-neutron gaps δ_{2n} (**top**) and quadrupole deformation β_2 (**bottom**) at the ground state for several isotopic chains, obtained from RHB calculations based on the DD-PC1 (**a**,**d**) and DD-ME2 (**b**,**e**) functionals, and from HFB calculations based on the UNEDF0 (**c**,**f**) functional. The hollow symbols indicate that the nucleus is unbound.

The results from calculations using the UNEDF0 functional exhibit significant differences compared to those with DD-ME2. As shown in Figure 1c, in the calculations spanning isotopic chains from $Z = 118$ to $Z = 132$, a broader but lower peak appears near $N = 258$, suggesting a weaker shell effect. The shape of the peak changes noticeably at $Z = 130$, and it disappears entirely in the $Z = 132$ chain. Upon examining the shape evolution in Figure 1f, similar to the case with DD-ME2, the competition between spherical and prolate shapes leads to the disappearance of the shell structure near $N = 258$. In Figure 1b,c, instances where the δ_{2n} becomes negative often correspond to sudden changes in the ground-state shape of the nucleus as illustrated in Figure 1e,f. These shape transitions cause the rate of change in the total energy along the isotopic chain to decrease and then increase with the addition of neutrons, resulting in negative values for the δ_{2n}. Since DD-PC1 produces unbound results, the following discussion will primarily focus on the results from DD-ME2 and UNEDF0.

Based on the DD-ME2 functional, the evolution of the spherical shell structure and density variation near $N = 258$ within isotopic chains $Z = 124$ and isotonic chains $N = 258$ is investigated, with the results presented in Figure 2. In the density distributions shown in Figure 2b,d, the solid lines represent neutron density, while the dash-dot lines represent proton density. As shown in Figure 2a, the spin–orbit splitting of the $1k$ state, along with the approximate restoration of pseudospin symmetry between the three pairs of pseudospin partners ($2h_{9/2}$, $3f_{7/2}$), ($3f_{5/2}$, $4p_{3/2}$) and ($1k_{15/2}$, $2i_{13/2}$) together contribute to the formation of large shell gaps at $N = 228$ and $N = 258$. In the isotopic chain, as the neutron number increases, the shell gap at $N = 258$ gradually decreases. This trend can be explained by the density variations observed in Figure 2b. As the neutron number increases from $N = 254$ to $N = 258$, the occupancy of the $4p$ orbital causes a sudden rise in neutron density at the center. As the neutron number continues to increase, the $1k_{15/2}$ orbital is occupied, but this only has a small impact on the shape of the potential, particularly on the radial profile at the bottom of the potential. The changes in the potential are reflected in the energy levels, and it can be seen that there are only limited changes in the energy levels near the Fermi surface in the isotopic chain.

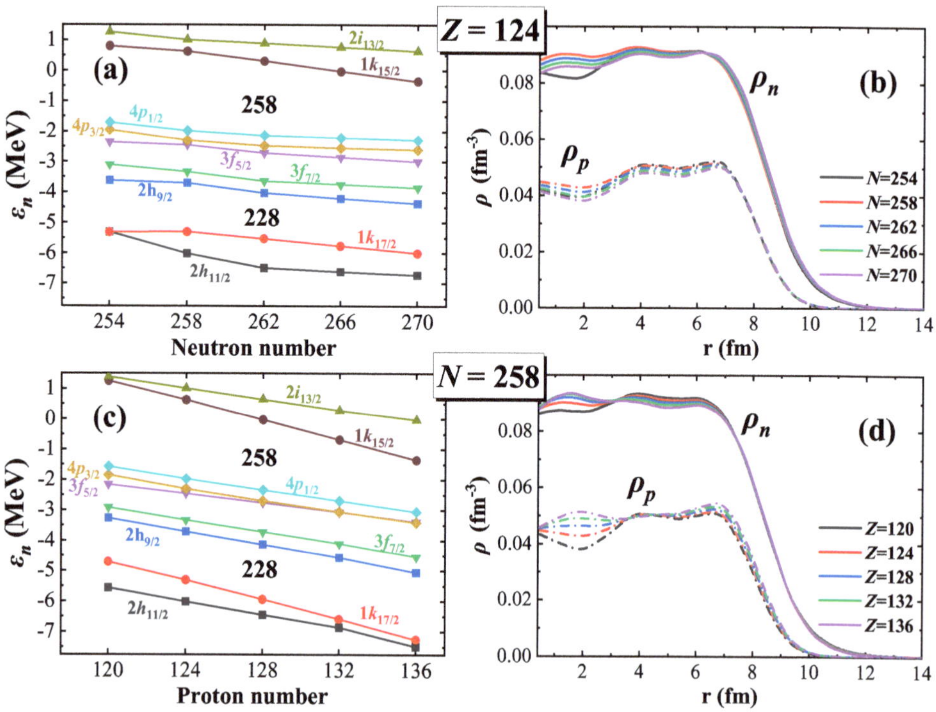

Figure 2. The neutron single-particle levels (**a**) and nucleon density distributions (**b**) for the spherical states of $Z = 124$ isotopes, as well as the neutron single-particle levels (**c**) and nucleon density distributions (**d**) for the spherical states of $N = 258$ isotones, obtained from RHB calculations based on the DD-ME2 functional. In the density distribution plots, the solid lines represent the neutron density distribution, while the dash-dot lines represent the proton density distribution.

For the isotonic chain $N = 258$ as shown in Figure 2c, the increase in proton number leads to a rapid decrease in the shell gap at $N = 258$. Observing the density evolution in Figure 2d, it can be observed that as the proton number increases, the occupation of low-j orbitals causes the central proton density to rise. In self-consistent calculations, the change in proton density feeds back to the neutrons through the vector potential, driving the density distributions of proton and neutron to become more similar. In the isotonic chain, this results in a significant change in the shape of the neutron potential. The alterations in the potential are reflected in the neutron energy levels, and it can be seen that as the proton number increases, the shell gap at $N = 258$ rapidly decreases. Ref. [37] has comprehensively discussed the physical mechanisms underlying the formation or suppression of the central density depression in nuclei.

To explain the differences observed near $N = 258$ with the UNEDF0 functional, Figure 3 presents the two-neutron gaps δ_{2n} calculated for the $Z = 124$ isotopic chain, both with and without the LN method. It can be observed that if the LN method is not employed to account for pairing, the trend of the δ_{2n} with the UNEDF0 functional near $N = 258$ becomes similar to the results from the DD-ME2 functional. Comparing the calculations with and without the LN method, the differences become more pronounced as one approaches $N = 258$. A similar result can also be observed in the calculations around $N = 184$.

Figure 4 shows the differences in neutron occupancy probabilities near the Fermi surface for the nuclei $^{380}124_{256}$ and $^{382}124_{258}$ under both the usage and non-usage of the LN method. In the case where the LN method is applied, comparing the results in Figure 4a,b, the neutron occupancy of $^{380}124_{256}$ is similar to that of $^{382}124_{258}$, with neutrons having a certain occupancy probability above the Fermi surface. Additionally, a significant change in the Fermi energy can be observed, and the shell gap at $N = 258$ is noticeably decreased. In contrast, without the LN method, there is a significant difference in neutron occupancy between $^{380}124_{256}$ and $^{382}124_{258}$. For $^{382}124_{258}$, the occupancy probability of levels above the Fermi surface is 0. This difference in occupancy probabilities feeds back into the density distribution, subsequently affecting the energy. The application of the LN method has a substantial impact on the kinetic energy, pairing energy, and volume energy, ultimately leading to smaller energy differences between the isotopes. This also explains why the δ_{2n} obtained with UNEDF0 in Figure 1c is generally smaller.

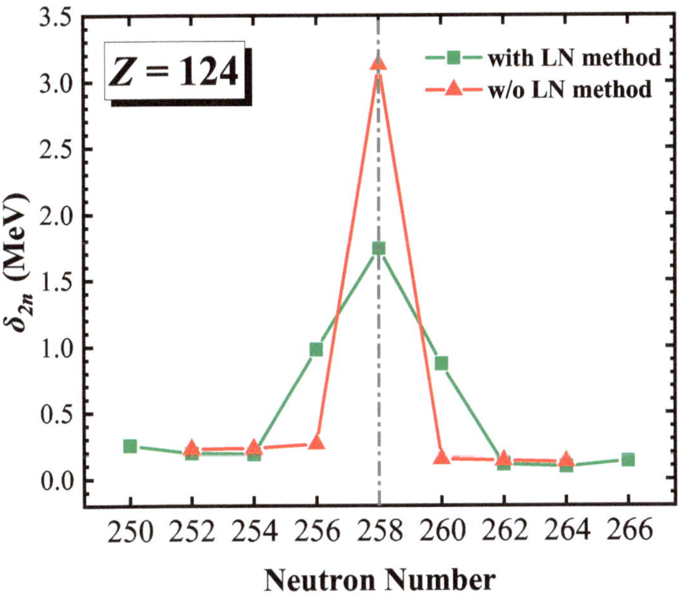

Figure 3. The two-neutron gaps δ_{2n} for the $Z = 124$ isotopic chain obtained from HFB calculations based on the UNEDF0 functional. Green squares indicate the use of the LN method to treat pairing, while red triangles represent calculations without the LN method.

Finally, based on the UNEDF0 functional, the evolution of the spherical shell structure and density distribution in the $Z = 124$ isotopic chain and the $N = 258$ isotonic chain are presented in Figure 5. In the density distributions shown in Figure 5b,d, the solid lines represent neutron density, while the dash-dot lines represent proton density. Figure 5a shows the changes in the spherical shell structure in the $Z = 124$ isotopic chain as the neutron number increases. Unlike the results from the DD-ME2 functional, the $1k_{15/2}$ and $2i_{13/2}$ orbitals have exchanged positions, and there is a higher degree of degeneracy between the $3f_{5/2}$ and $4p_{3/2}$, $2h_{9/2}$ and $3f_{7/2}$, as well as between the $1j_{13/2}$ and $2h_{11/2}$. Due to the use of the LN method, the energy level spacing between $4p_{1/2}$ and $2i_{13/2}$ at $N = 258$ decreases in the isotopic chain, which is consistent with the results shown in Figure 4. Furthermore, the shell gap at $N = 258$ does not decrease with increasing the neutron number but remains stable. Observing the density evolution presented in Figure 5b, as previously mentioned, the neutron and proton density distributions show no significant changes due to the use of the LN method, which corresponds to the stability of the shell structure.

For the isotonic chain at $N = 258$, Figure 5c shows that as the proton number increases, the neutron shell gap caused by the $4p_{1/2}$ and $2i_{13/2}$ orbitals remains essentially unchanged. However, the energy of the $1k$ orbital drops at a faster rate. It can be anticipated that as the proton number continues to increase, the spherical shell gap at $N = 258$ will be determined by the $4p_{1/2}$ and $1k_{15/2}$ orbitals and will gradually decrease.

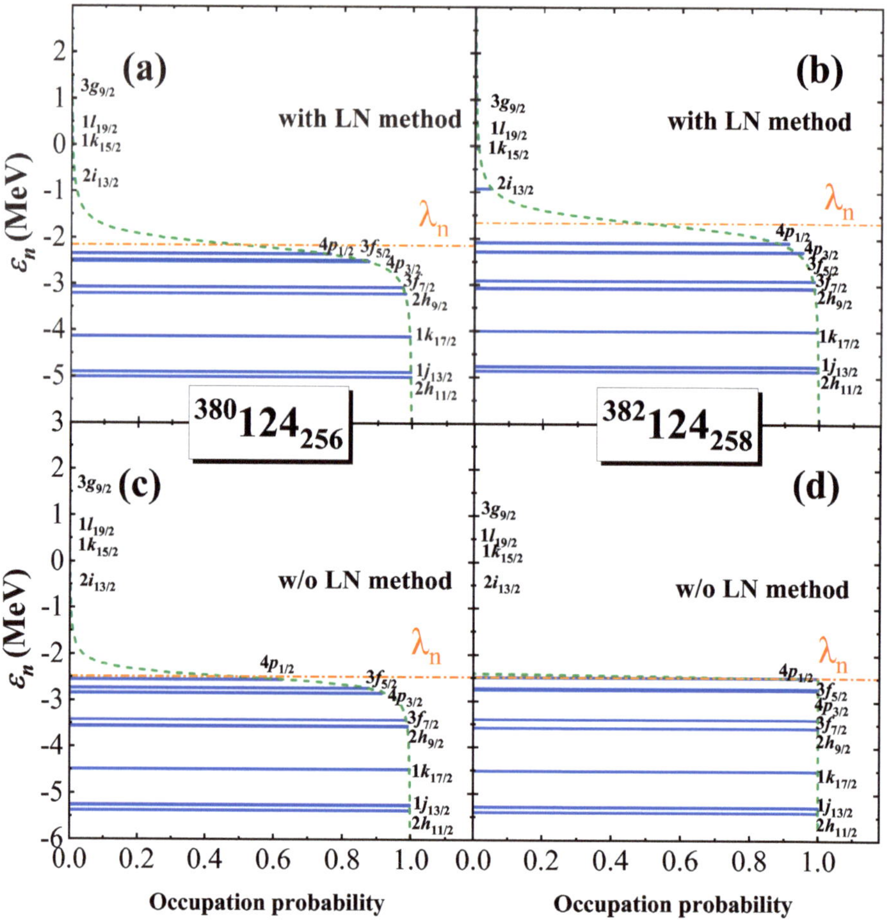

Figure 4. The single-neutron level occupation probabilities for $^{380}124_{256}$ (**a**,**c**) and $^{382}124_{258}$ (**b**,**d**) obtained from HFB calculations based on the UNEDF0 functional. The top panels use the LN method to treat pairing, while the bottom panels do not. The yellow dash-dot line represents the neutron Fermi energy. The green dashed line corresponds to the BCS formula with an average pairing gap.

Figure 5d shows the evolution of the density distribution. As the proton number increases, the central proton density grows, but unlike the results from DD-ME2, this change does not significantly impact the neutron density distribution. However, a slight decrease in neutron density near the surface is still observable, indicating that the density distribution of high-j orbitals is increasingly concentrated toward the center. This is consistent with the evolution of the energy levels.

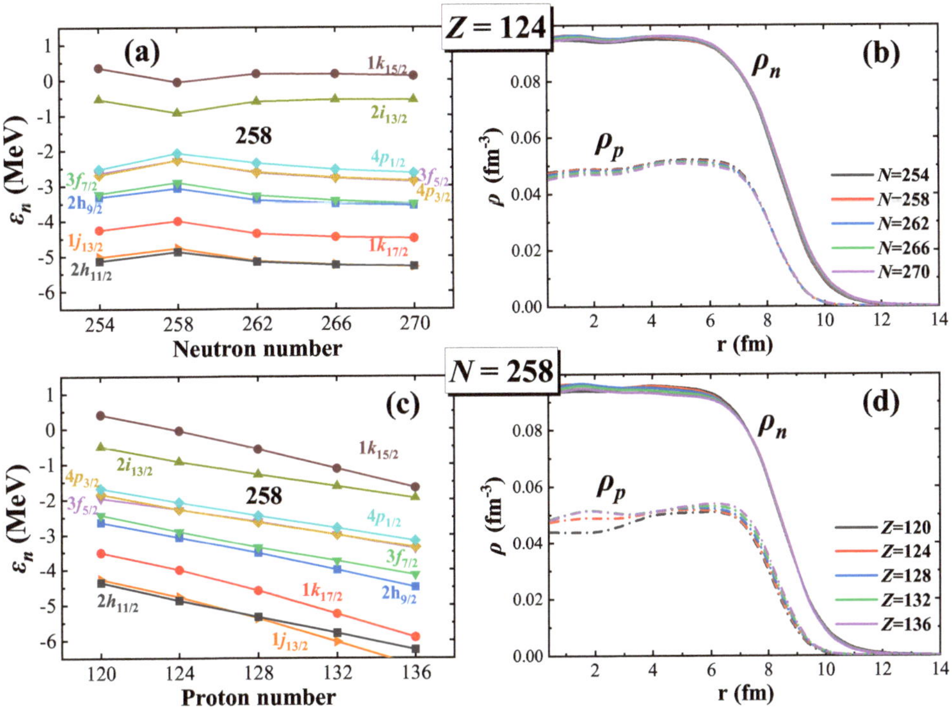

Figure 5. The neutron single-particle levels (**a**) and nucleon density distributions (**b**) for the spherical states of $Z = 124$ isotopes, as well as the neutron single-particle levels (**c**) and nucleon density distributions (**d**) for the spherical states of $N = 258$ isotones, obtained from the HFB calculations based on the UNEDF0 functional. In the density distribution plots, the solid lines represent the neutron density distribution, while the dash-dot lines represent the proton density distribution.

4. Conclusions

Based on relativistic functionals DD-PC1 and DD-ME2, and the non-relativistic functional UNEDF0, calculations for superheavy nuclei were performed within the axially symmetric RHB and HFB frameworks, respectively. The DD-PC1 functional did not predict a bound magic nucleus at $N = 258$, but the results indicated a significant shell gap at this neutron number. Calculations using the DD-ME2 functional yielded a sharp and narrow peak in the two-neutron gap at $N = 258$, indicating the presence of a magic number. In contrast, the UNEDF0 functional produced a wider and lower peak for the two-neutron gap.

Further comparison of HFB calculations based on the UNEDF0 functional, with and without the LN method, revealed that the use of the LN method caused the $N = 258$ nucleus to have similar single-particle orbital occupations to its neighboring nuclei. This led to similar density distributions and energies across isotopes, resulting in a significantly smaller two-neutron gap compared to the results from DD-ME2.

Due to the competition between spherical and prolate deformations, the neutron magic number $N = 258$ disappears at $Z = 126$ for DD-ME2 and at $Z = 132$ for UNEDF0. By examining the evolution of the spherical shell structure and the nucleon density distribution within isotopic and isotonic chains, it was found that a decreasing central density is more favorable for maintaining the magic number $N = 258$. A similar conclusion was observed for $Z = 120$ and $N = 172$ [18]. For the neutron magic number $N = 258$, the results indicate that the central density depression is primarily driven by the proton contribution.

In summary, both the DD-ME2 and UNEDF0 functionals predict that $N = 258$ is a neutron magic number, and the emergence of such a shell structure is related to the depression of the central density. However, this does not necessarily imply the existence of a superheavy stability island near $N = 258$. Further triaxial deformation calculations are needed to assess whether the nuclei in this region will be reasonably stable against spontaneous fission.

Author Contributions: Conceptualization, P.D. and J.L.; formal analysis, P.D.; investigation, P.D.; resources, J.L.; data curation, P.D.; writing—original draft preparation, P.D.; writing—review and editing, P.D. and J.L.; visualization, P.D.; supervision, J.L.; project administration, J.L.; funding acquisition, J.L. All authors have read and agreed to the published version of the manuscript.

Funding: This research was funded by the National Natural Science Foundation of China (Nos. 12475119 and 11675063), Scientific Research Project of Education Department of Jilin Province (No. JJKH20241242KJ), Natural Science Foundation of Jilin Province (No.20220101017JC), and Key Laboratory of Nuclear Data Foundation (JCKY2020201C157).

Data Availability Statement: Data will be made available on request.

Conflicts of Interest: The funders had no role in the design of the study; in the collection, analyses, or interpretation of data; in the writing of the manuscript; or in the decision to publish the results.

References

1. Oganessian, Y.T.; Utyonkov, V.K.; Lobanov, Y.V.; Abdullin, F.S.; Polyakov, A.N.; Sagaidak, R.N.; Shirokovsky, I.V.; Tsyganov, Y.S.; Voinov, A.A.; Gulbekian, G.G.; et al. Synthesis of the isotopes of elements 118 and 116 in the ^{249}Cf and ^{245}Cm $+^{48}$ Ca fusion reactions. *Phys. Rev. C* **2006**, *74*, 044602. [CrossRef]
2. Oganessian, Y.T.; Abdullin, F.S.; Alexander, C.; Binder, J.; Boll, R.A.; Dmitriev, S.N.; Ezold, J.; Felker, K.; Gostic, J.M.; Grzywacz, R.K.; et al. Production and Decay of the Heaviest Nuclei ^{293,294}117 and ^{294}118. *Phys. Rev. Lett.* **2012**, *109*, 162501. [CrossRef] [PubMed]
3. Oganessian, Y.T.; Utyonkov, V.K. Super-heavy element research. *Rep. Prog. Phys.* **2015**, *78*, 036301. [CrossRef] [PubMed]
4. Oganessian, Y.T.; Utyonkov, V.K.; Lobanov, Y.V.; Abdullin, F.S.; Polyakov, A.N.; Sagaidak, R.N.; Shirokovsky, I.V.; Tsyganov, Y.S.; Voinov, A.A.; Mezentsev, A.N.; et al. Attempt to produce element 120 in the ^{244}Pu $+^{58}$ Fe reaction. *Phys. Rev. C* **2009**, *79*, 024603. [CrossRef]
5. Khuyagbaatar, J.; Yakushev, A.; Düllmann, C.E.; Ackermann, D.; Andersson, L.L.; Asai, M.; Block, M.; Boll, R.A.; Brand, H.; Cox, D.M.; et al. Search for elements 119 and 120. *Phys. Rev. C* **2020**, *102*, 064602. [CrossRef]
6. Sakai, H.; Haba, H.; Morimoto, K.; Sakamoto, N. Facility upgrade for superheavy-element research at RIKEN. *Eur. Phys. J. A* **2022**, *58*, 238. [CrossRef]
7. Gan, Z.G.; Huang, W.X.; Zhang, Z.Y.; Zhou, X.H.; Xu, H.S. Results and perspectives for study of heavy and super-heavy nuclei and elements at IMP/CAS. *Eur. Phys. J. A* **2022**, *58*, 158. [CrossRef]
8. Meng, J.; Toki, H.; Zhou, S.G.; Zhang, S.Q.; Long, W.H.; Geng, L.S. Relativistic continuum Hartree Bogoliubov theory for ground-state properties of exotic nuclei. *Prog. Part. Nucl. Phys.* **2006**, *57*, 470–563. [CrossRef]
9. Nikšić, T.; Vretenar, D.; Ring, P. Relativistic nuclear energy density functionals: Mean-field and beyond. *Prog. Part. Nucl. Phys.* **2011**, *66*, 519–548. [CrossRef]
10. Bartel, J.; Quentin, P.; Brack, M.; Guet, C.; Håkansson, H.-B. Towards a better parametrisation of Skyrme-like effective forces: A critical study of the SkM force. *Nucl. Phys. A* **1982**, *386*, 79–100. [CrossRef]
11. Chabanat, E.; Bonche, P.; Haensel, P.; Meyer, J.; Schaeffer, R. A Skyrme parametrization from subnuclear to neutron star densities Part II. Nuclei far from stabilities. *Nucl. Phys. A* **1998**, *635*, 231–256. [CrossRef]
12. Robledo, L.M.; Rodríguez, T.R.; Rodríguez-Guzmán, R.R. Mean field and beyond description of nuclear structure with the Gogny force: A review. *J. Phys. G Nucl. Part. Phys.* **2018**, *46*, 013001. [CrossRef]
13. Vautherin, D.; Brink, D.M. Hartree-Fock Calculations with Skyrme's Interaction. I. Spherical Nuclei. *Phys. Rev. C* **1972**, *5*, 626–647. [CrossRef]
14. Rutz, K.; Bender, M.; Bürvenich, T.; Schilling, T.; Reinhard, P.-G.; Maruhn, J.A.; Greiner, W. Superheavy nuclei in self-consistent nuclear calculations. *Phys. Rev. C* **2008**, *56*, 238–243. [CrossRef]
15. Bender, M.; Nazarewicz, W.; Reinhard, P.-G. Shell stabilization of super- and hyperheavy nuclei without magic gaps. *Phys. Lett. B* **2001**, *515*, 42–48. [CrossRef]
16. Zhang, W.; Meng, J.; Zhang, S.Q.; Geng, L.S.; Toki, H. Magic numbers for superheavy nuclei in relativistic continuum Hartree–Bogoliubov theory. *Nucl. Phys. A* **2005**, *753*, 106–135. [CrossRef]
17. Li, J.J.; Long, W.H.; Margueron, J.; Van Giai, N. Superheavy magic structures in the relativistic Hartree–Fock–Bogoliubov approach. *Phys. Lett. B* **2014**, *732*, 169–173. [CrossRef]
18. Afanasjev, A.V.; Frauendorf, S. Central depression in nuclear density and its consequences for the shell structure of superheavy nuclei. *Phys. Rev. C* **2005**, *71*, 024308. [CrossRef]

19. Agbemava, S.E.; Afanasjev, A.V.; Nakatsukasa, T.; Ring, P. Covariant density functional theory: Reexamining the structure of superheavy nuclei. *Phys. Rev. C* **2015**, *92*, 054310. [CrossRef]
20. Afanasjev, A.V.; Agbemava, S.E.; Gyawali, A. Hyperheavy nuclei: Existence and stability. *Phys. Lett. B* **2018**, *782*, 533–540. [CrossRef]
21. Taninah, A.; Agbemava, S.E.; Afanasjev, A.V. Covariant density functional theory input for *r*-process simulations in actinides and superheavy nuclei: The ground state and fission properties. *Phys. Rev. C* **2020**, *102*, 054330. [CrossRef]
22. Zhao, P.W.; Li, Z.P.; Yao, J.M.; Meng, J. New parametrization for the nuclear covariant energy density functional with a point-coupling interaction. *Phys. Rev. C* **2010**, *82*, 054319. [CrossRef]
23. Li, L.L.; Meng, J.; Ring, P.; Zhao, E.-G.; Zhou, S.-G. Deformed relativistic Hartree-Bogoliubov theory in continuum. *Phys. Rev. C* **2012**, *85*, 024312. [CrossRef]
24. Zhang, K.; Cheoun, M.K.; Choi, Y.B.; Chong, P.S.; Dong, J.; Dong, Z.; Du, X.; Geng, L.; Ha, E.; He, X.T.; et al. Nuclear mass table in deformed relativistic Hartree–Bogoliubov theory in continuum, I: Even–even nuclei. *At. Data Nucl. Data Tables* **2022**, *144*, 101488. [CrossRef]
25. Guo, P.; Cao, X.; Chen, K.; Chen, Z.; Cheoun, M.K.; Choi, Y.B.; Lam, P.C.; Deng, W.; Dong, J.; Du, P.; et al. Nuclear mass table in deformed relativistic Hartree–Bogoliubov theory in continuum, II: Even-Z nuclei. *At. Data Nucl. Data Tables* **2024**, *158*, 101661. [CrossRef]
26. Zhang, Y.X.; Liu, B.R.; Zhang, K.Y.; Yao, J.M. Shell structure and shape transition in odd-Z superheavy nuclei with proton numbers $Z = 117, 119$: Insights from applying deformed relativistic Hartree-Bogoliubov theory in continuum. *Phys. Rev. C* **2024**, *110*, 024302. [CrossRef]
27. Nikšić, T.; Vretenar, D.; Ring, P. Relativistic nuclear energy density functionals: Adjusting parameters to binding energies. *Phys. Rev. C* **2008**, *78*, 034318. [CrossRef]
28. Lalazissis, G.A.; Nikšić, T.; Vretenar, D.; Ring, P. New relativistic mean-field interaction with density-dependent meson-nucleon couplings. *Phys. Rev. C* **2005**, *71*, 024312. [CrossRef]
29. Kortelainen, M.; Lesinski, T.; Moré, J.; Nazarewicz, W.; Sarich, J.; Schunck, N.; Stoitsov, M.V.; Wild, S. Nuclear energy density optimization. *Phys. Rev. C* **2010**, *82*, 024313. [CrossRef]
30. Nikšić, T.; Paar, N.; Vretenar, D.; Ring, P. DIRHB—A relativistic self-consistent mean-field framework for atomic nuclei. *Comput. Phys. Commun.* **2014**, *185*, 1808–1821. [CrossRef]
31. Tian, Y.; Ma, Z.Y.; Ring, P. A finite range pairing force for density functional theory in superfluid nuclei. *Phys. Lett. B* **2009**, *676*, 44–50. [CrossRef]
32. Karatzikos, S.; Afanasjev, A.V.; Lalazissis, G.A.; Ring, P.T. The fission barriers in Actinides and superheavy nuclei in covariant density functional theory. *Phys. Lett. B* **2010**, *689*, 72–81. [CrossRef]
33. Stoitsov, M.V.; Dobaczewski, J.; Nazarewicz, W.; Ring, P. Axially deformed solution of the Skyrme–Hartree–Fock–Bogolyubov equations using the transformed harmonic oscillator basis. The program HFBTHO (v1.66p). *Comput. Phys. Commun.* **2005**, *167*, 43–63. [CrossRef]
34. Marević, P.; Schunck, N.; Ney, E.M.; Navarro Pérez, R.; Verriere, M.; O'Neal, J. Axially-deformed solution of the Skyrme-Hartree-Fock-Bogoliubov equations using the transformed harmonic oscillator basis (IV) hfbtho (v4.0): A new version of the program. *Comput. Phys. Commun.* **2022**, *276*, 108367. [CrossRef]
35. Reinhard, P.-G.; Nazarewicz, W.; Bender, M.; Maruhn, J.A. Lipkin-Nogami pairing scheme in self-consistent nuclear structure calculations. *Phys. Rev. C* **1996**, *53*, 2776–2785. [CrossRef]
36. Stoitsov, M.V.; Dobaczewski, J.; Nazarewicz, W.; Pittel, S.; Dean, D.J. Systematic study of deformed nuclei at the drip lines and beyond. *Phys. Rev. C* **2003**, *68*, 054312. [CrossRef]
37. Perera, U.C.; Afanasjev, A.V. Bubble nuclei: Single-particle versus Coulomb interaction effects. *Phys. Rev. C* **2022**, *106*, 024321. [CrossRef]

particles

MDPI

Article

Magic Number $N = 350$ Predicted by the Deformed Relativistic Hartree-Bogoliubov Theory in Continuum: $Z = 136$ Isotopes as an Example

Wei-Jian Liu [1], Chen-Jun Lv [1], Peng Guo [2], Cong Pan [3,4], Sibo Wang [5,6] and Xin-Hui Wu [1,2,*]

1 Department of Physics, Fuzhou University, Fuzhou 350108, China; 052203115@fzu.edu.cn (W.-J.L.)
2 State Key Laboratory of Nuclear Physics and Technology, School of Physics, Peking University, Beijing 100871, China; 2301110125@pku.edu.cn
3 Department of Physics, Anhui Normal University, Wuhu 241000, China; cpan@ahnu.edu.cn
4 Center for Exotic Nuclear Studies, Institute for Basic Science, Daejeon 34126, Republic of Korea
5 Department of Physics, Chongqing University, Chongqing 401331, China
6 Chongqing Key Laboratory for Strongly Coupled Physics, Chongqing 401331, China
* Correspondence: wuxinhui@fzu.edu.cn

Abstract: The investigation of magic numbers for nuclei in the hyperheavy region ($Z > 120$) is an interesting topic. The neutron magic number $N = 350$ is carefully validated by the deformed relativistic Hartree-Bogoliubov theory in continuum (DRHBc), via analysing even-even nuclei around $N = 350$ of the $Z = 136$ isotopes in detail. Nuclei with $Z = 136$ and $340 \leq N \leq 360$ are all found to be spherical in their ground states. A big drop of the two-neutron separation energy S_{2n} is observed from $N = 350$ to $N = 352$ in the isotopic chain of $Z = 136$, and a peak of the two-neutron gap δ_{2n} appears at $N = 350$. There exists a big shell gap above $N = 350$ around the spherical regions of single-neutron levels for nucleus with ($Z = 136, N = 350$). These evidences from the DRHBc theory support $N = 350$ to be a neutron magic number in the hyperheavy region.

Keywords: relativistic density functional theory; deformed relativistic Hartree-Bogoliubov theory in continuum; magic number; hyperheavy nuclei

1. Introduction

The investigation of superheavy elements remains one of the most important topics of nuclear physics and chemistry. The element with the largest proton number Z observed so far is Og with $Z = 118$ [1]. The limit of the existence of atomic nuclei is a longstanding issue for both experimental and theoretical nuclear physicists, and has important impacts on physics and chemistry. The nuclei with $Z > 120$ are usually called hyperheavy nuclei [2]. The studies of hyperheavy nuclei can enhance our understanding of exotic nuclear structures and enable the delving into the limits of charge and mass of atomic nuclei.

Nuclear liquid drop model (LDM) [3] can help us obtain a quick understanding of the hyperheavy nuclei, which suggests the importance of the competition between Coulomb energies and surface effects in the hyperheavy region. However, due to the lack of quantum shell effect, the predictions given by the LDM are pretty rough. Quantum shell effect, which corresponds to a non-uniformity distribution of the individual single-particle energies, is very important for finite nuclear systems. It can produce a significant energy gap in the single-particle energy spectrum near the Fermi level for some nuclei. Such gaps would provide additional binding energies and enhance nuclear stability. These nuclei with additional stability are the so-called "magic nuclei", and the corresponding proton or neutron numbers are called "magic numbers". Experimentally, the confirmed neutron magic numbers are $8, 20, 28, 50, 82, 126$, and the confirmed proton magic numbers are $8, 20, 28, 50, 82$.

Due to the additional stability, magic nuclei have drawn a lot of attention [4,5]. Naïvely, if one assumes the potential for nucleons within an atomic nucleus is a harmonic oscillator potential, the obtained magic numbers will be $2, 8, 20, 40, 70, 112, 168, 240, \ldots$, which disagree with experiments. If one further takes into account the spin-orbital coupling, the predicted magic numbers are $2, 8, 20, 28, 50, 82, 126, 184, 258, 350, \ldots$, which correctly reproduce the experimental magic numbers for $N \leq 126$ and $Z \leq 82$. However, due to the limit of experimental information, the large predicted magic numbers, such as 258 and 350, are difficult to be validated in the foreseeable future. For hyperheavy nuclei with $Z > 120$, the neutron numbers of which can reach to $N \approx 350$, and it will be interesting to use a microscopic and self-consistent model to theoretically justify the predicted neutron magic number 350.

Nuclear stability is usually described by the binding energy or equivalently nuclear mass. Nuclear mass is important for both nuclear physics [6,7] and astrophysics [8–11]. Experimentally, the masses of about 2500 nuclear masses have been measured to date [12]. Theoretically, many nuclear models [13–24] and machine-learning approaches [25–36] are developed to predict nuclear masses. Among these models, the deformed relativistic Hartree-Bogoliubov theory in continuum (DRHBc) [37,38] simultaneously treats the deformation degrees of freedom, pairing correlations, and continuum effects properly, which are important for the descriptions of weakly bound exotic nuclei. The DRHBc theory has been successfully applied in studying many nuclear phenomena [39–56]. In order to provide a unified and microscopic description for the whole nuclear landscape, the DRHBc Mass Table Collaboration [57] was established, aiming at establishing a nuclear mass table based on the DRHBc theory with the density functional PC-PK1 [58]. The even-even [22] and even-odd [24] parts of the DRHBc mass table have been established recently. The Collaboration is now working on odd-Z nuclei and hyperheavy nuclei with $120 < Z \leq 136$. Taking this opportunity, one can validate the neutron magic number $N = 350$ with the DRHBc theory. In this work, the DRHBc theory is employed to study the even nuclei of $Z = 136$ isotopes around $N = 350$ to validate the possible neutron magic number $N = 350$. To our knowledge, there is currently no literature that employs modern nuclear model to study the neutron magic number $N = 350$. In Section 2, the theoretical framework is introduced. The numerical details are introduced in Section 3. The Results and discussions are presented in Section 4. Finally, a summary is given in Section 5.

2. Theoretical Framework

The details of the DRHBc theory with meson-exchange and point-coupling density functionals can be found in Refs. [38,59], respectively. In the following we briefly present its formalism.

Treating self-consistently the mean fields and pairing correlations, the relativistic Hartree Bogoliubov (RHB) equations for the nucleons read [60]

$$\begin{pmatrix} h_D - \lambda_\tau & \Delta \\ -\Delta^* & -h_D^* + \lambda_\tau \end{pmatrix} \begin{pmatrix} U_k \\ V_k \end{pmatrix} = E_k \begin{pmatrix} U_k \\ V_k \end{pmatrix}. \tag{1}$$

To describe properly the possible large spatial extension of exotic nuclei, the RHB equations are solved in a Dirac Woods-Saxon basis, in which the radial wave functions have a proper asymptotic behavior for large r [61]. In Equation (1), λ_τ is the Fermi energy ($\tau = n/p$ for neutrons or protons), E_k and $(U_k, V_k)^{\mathrm{T}}$ the quasiparticle energy and wave function, and h_D the Dirac Hamiltonian,

$$h_D(\boldsymbol{r}) = \boldsymbol{\alpha} \cdot \boldsymbol{p} + V(\boldsymbol{r}) + \beta[M + S(\boldsymbol{r})], \tag{2}$$

with the scalar $S(\boldsymbol{r})$ and vector $V(\boldsymbol{r})$ potentials. The pairing potential Δ reads

$$\Delta(\boldsymbol{r}_1, \boldsymbol{r}_2) = V^{\mathrm{pp}}(\boldsymbol{r}_1, \boldsymbol{r}_2)\kappa(\boldsymbol{r}_1, \boldsymbol{r}_2), \tag{3}$$

with a density-dependent force of zero range,

$$V^{\rm pp}(\boldsymbol{r}_1,\boldsymbol{r}_2) = V_0\frac{1}{2}(1-P^{\sigma})\delta(\boldsymbol{r}_1-\boldsymbol{r}_2)\left(1-\frac{\rho(\boldsymbol{r}_1)}{\rho_{\rm sat}}\right), \tag{4}$$

and the pairing tensor κ [62]. For axially deformed nuclei, the potentials and densities are expanded in terms of the Legendre polynomials,

$$f(\boldsymbol{r}) = \sum_{\lambda} f_{\lambda}(r)P_{\lambda}(\cos\theta), \ \ \lambda = 0,2,4,\cdots, \tag{5}$$

where λ is restricted to be even numbers due to spatial reflection symmetry.

3. Numerical Details

In Equation (4), the pairing strength $V_0 = -325$ MeV fm^3, the saturation density $\rho_{\rm sat} = 0.152$ fm^{-3}, and a pairing window of 100 MeV are adopted. The energy cutoff $E^+_{\rm cut}$ = 300 MeV and the angular momentum cutoff $J_{\rm max}$ = 23/2 $\hbar$ are adopted for the Dirac Woods-Saxon basis. In Equation (5), the Legendre expansion is truncated at $\lambda_{\rm max} = 10$ [63]. The calculations are carried out with the relativistic density functional PC-PK1 [58]. These numerical details are the same as the calculations of nuclei with $100 \leq Z \leq 120$ in the global DRHBc mass table calculations over the nuclear chart [22,24,59], and have also been examined to be proper for the studies in this work.

4. Results and Discussions

Evolution of the potential energy curves (PECs) of $Z = 136$ isotopes with $340 \leq N \leq 360$ is presented in Figure 1. Note that the $Z = 136$ element has temporary systematic IUPAC name and symbol as Untrihexium and Uth respectively [64]. One can see the similar behaviours of these PECs, where the total energy increases monotonously with the increasing absolute value of β_2 in the range that $|\beta_2| < 0.3$, indicating that these nuclei are spherical in their ground states. Note that the cutoff of angular momentum has been examined in Ref. [65], which suggests $J_{max} = 31/2\,\hbar$ in the calculation of the hyperheavy nuclei ($121 \leq Z \leq 136$) with quadruple deformation $|\beta_2| > 0.3$. The spherical ground states of these nuclei can be interpreted as a clue that $N = 350$ (or other adjacent neutron numbers) is a neutron magic number, since atomic nuclei prefer to be spherical around the magic ones. Note that the nuclei that are analyzed in the present work all have spherical ground states, which might lead one to consider $Z = 136$ as a potential magic number. However, we have checked the nuclei that are not close to $N = 350$ in $Z = 136$ isotopic chain, most of which are not spherical ones. We have also checked the single-proton levels around the Fermi level of ^{486}Uth , and there is not a significant gap above $Z = 136$. Instead, there is a significant gap above $Z = 138$, which may suggest 138 to be a proton magic number. However, this is out of the scope of the present paper, which could be an interesting topic for the future work.

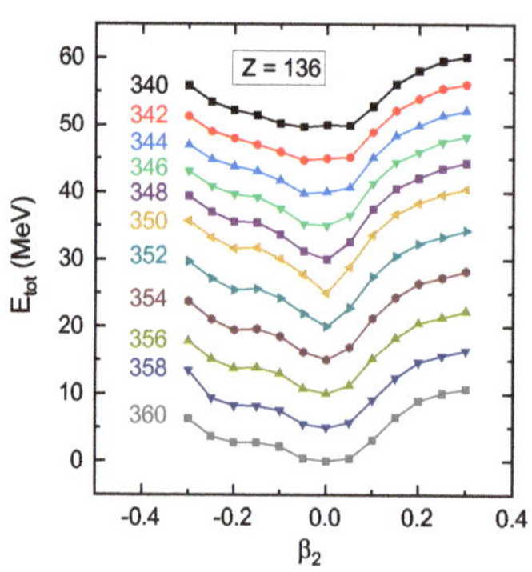

Figure 1. Evolution of the potential energy curves (PECs) of Uth ($Z = 136$) isotopes with $340 \leq N \leq 360$. For clarity reasons, the curves have been scaled to the energy of $\beta_2 = 0$ and have been shifted upward by 5 MeV per decreasing two neutrons.

In order to pin down which neutron number is the magic one, the two-neutron separation energies S_{2n} and Fermi energies λ_n for these Uth isotopes are presented in Figure 2. One can see that the two-neutron separation energies S_{2n} evolve slowly with the increasing of neutron number from $N = 340$ to $N = 350$ and from $N = 352$ to $N = 360$. However, there is a big drop in the S_{2n} from $N = 350$ to $N = 352$. This indicates that there is a big shell gap at $N = 350$, which is a strong evidence for a magic number. There is also a big jump in the Fermi energies λ_n from $N = 350$ to $N = 352$, which leads to the same conclusion. Two-neutron gap $\delta_{2n} = S_{2n}(N, Z) - S_{2n}(N + 2, Z)$ is also a very good signature of magic numbers. As can be clearly seen in Figure 2, a peak of δ_{2n} appears at $N = 350$.

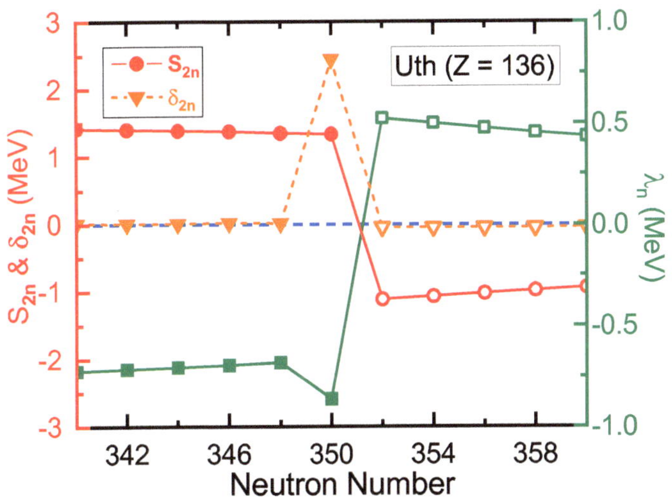

Figure 2. Two-neutron separation energies S_{2n}, two-neutron gaps δ_{2n}, and Fermi energies λ_n for the Uth ($Z = 136$) isotopes with $340 \leq N \leq 360$. The bound nuclei predicted by the DRHBc theory are denoted by filled circles, while the unbound nuclei are denoted by empty circles. The blue dashed line displays $S_{2n} = 0$, $\delta_{2n} = 0$, and $\lambda_n = 0$.

Neutron magic nuclei are typically more stable than their next neutron-rich neighbors, while the neutron-richer nuclei next to the magic nuclei are much more likely to emit extra neutrons outside the shells. One can also see in Figure 2 that the S_{2n} of nuclei $N \geq 352$ are smaller than zero, which means that these nuclei are unstable against neutron emission. The corresponding Fermi energies for these nuclei are larger than 0, which also refers to the unstable characters. Therefore, the neutron drip-line nucleus of Uth isotopic chain locates at $N = 350$. Note that the Coulomb repulsion will be very large in superheavy and hyperheavy nuclei, and tends to prevent the nuclear binding. The reason why such a hyperheavy nucleus ^{486}Uth still can be bound is due to strong shell effect. The shell effect is a hallmark characteristic in the atomic nucleus as a quantum system, providing extra binding that can overcome Coulomb repulsion and makes the nucleus bound.

In order to confirm the neutron magic number, the single-neutron levels around the Fermi level should be carefully checked. Figure 3 shows the single-neutron levels around the Fermi level of ^{486}Uth in the canonical basis obtained from constraint calculations. One can find a big gap above $N = 350$ around the spherical regions, i.e., $-0.05 < \beta_2 < 0.05$. This strongly supports $N = 350$ as a neutron magic number.

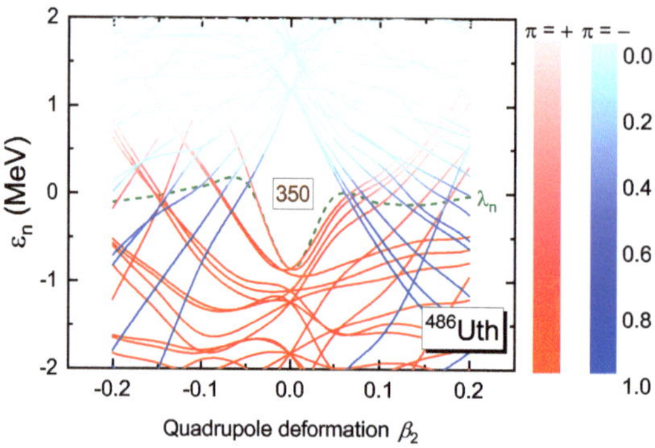

Figure 3. Single-neutron levels around the Fermi level of ^{486}Uth in the canonical basis obtained from constraint calculations with the DRHBc theory. The occupation probability of each orbital is represented with different colors. The Fermi level λ_n is displayed by the green dashed line.

Figure 4 provides single-neutron and single-proton levels around the Fermi levels for the spherical ground state of ^{486}Uth, and the corresponding spherical quantum numbers are labeled. As can be seen, the big gap of $N = 350$ appears between the level $4\mathrm{d}_{3/2}$ and $1\mathrm{I}_{17/2}$. In the future works, it would be interesting to check the model dependence of neutron magic number $N = 350$, as well as the related levels, among different methods and different functionals employed.

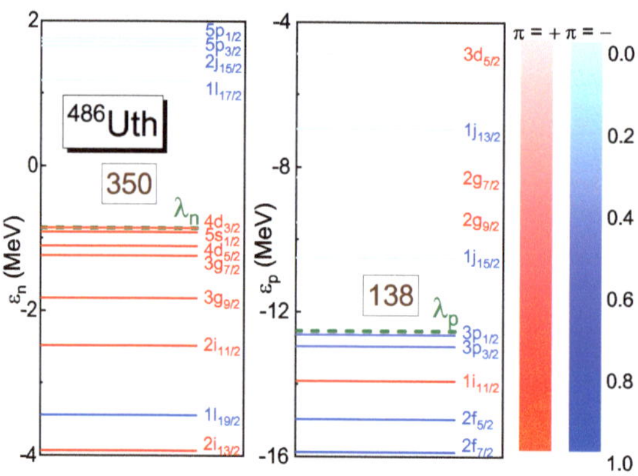

Figure 4. Single-neutron and single-proton levels around the Fermi levels for the spherical ground state of ^{486}Uth in the canonical basis obtained with the DRHBc theory. The occupation probability of each orbital is represented with different colors. The Fermi levels λ_n, λ_p are displayed by the green dashed lines. The spherical quantum numbers are given for corresponding levels.

5. Summary

The investigation of magic numbers of hyperheavy nuclei is an interesting topic. The neutron number $N = 350$ is predicted to be a magic number by the naïve analysis based on harmonic oscillator potential with spin-orbital coupling. In this work, the predicted neutron magic number $N = 350$ is validated with the DRHBc theory by studying the Uth ($Z = 136$) isotopes around $N = 350$. It is found that the Uth isotopes with $340 \leq N \leq 360$ are all spherical in their ground states. A big drop of the S_{2n} appears from $N = 350$ to $N = 352$, and a peak of δ_{2n} is observed at $N = 350$. By taking ^{486}Uth as an example, in the single-neutron levels, there exists a big shell gap above $N = 350$ around the spherical regions, i.e., $-0.05 < \beta_2 < 0.05$. These evidences from DRHBc theory all support $N = 350$ to be a neutron magic number in the hyperheavy region.

Note that we presently focus on $Z = 136$ isotopic chain as an example, and a more rigorous validation of neutron magic number should be examined also for other isotopic chains. The preliminary results of other isotopic chains in the hyperheavy region from the DRHBc theory also support $N = 350$ to be a neutron magic number. A more comprehensive investigation could be carried out in the future to analyze all the results of hyperheavy nuclei. In that case, one could have a more rigorous validation of neutron magic number $N = 350$ and also other possible neutron and proton magic numbers in the hyperheavy region. Considering that the present calculations with the DRHBc theory assume nuclei to be axial symmetry and base on the PC-PK1 functional, it would be interesting to investigate the evolution of shell gaps with triaxial deformation or with octupole deformation, to validate whether the triaxial or octupole deformations can challenge the spherical minima of these nuclei. It would also be interesting to check the functional dependence of neutron magic number $N = 350$.

Author Contributions: Conceptualization, X.-H.W.; methodology, C.P. and P.G.; formal analysis, C.P., P.G. and S.W.; investigation, W.-J.L. and C.-J.L.; writing—original draft preparation, X.-H.W.; writing—review and editing, All authors; visualization, W.-J.L., C.-J.L. and X.-H.W.; supervision, X.-H.W.; project administration, X.-H.W.; funding acquisition, S.W. and X.-H.W. All authors have read and agreed to the published version of the manuscript.

Funding: This work was partly supported by the National Natural Science Foundation of China under Grant No. 12405134 and No. 12205030, the State Key Laboratory of Nuclear Physics and Technology, Peking University under Grant No. NPT2023KFY02, the China Postdoctoral Science Foundation under Grant No. 2021M700256, and the start-up grant XRC-23103 of Fuzhou University.

Data Availability Statement: The dataset can be accessed upon request to the corresponding author.

Acknowledgments: Helpful discussions with members of the DRHBc Mass Table Collaboration are highly appreciated.

Conflicts of Interest: The authors declare no conflicts of interest.

References

1. Oganessian, Y.T.; Abdullin, F.S.; Alexander, C.; Binder, J.; Boll, R.A.; Dmitriev, S.N.; Ezold, J.; Felker, K.; Gostic, J.M.; Grzywacz, R.K.; et al. Production and Decay of the Heaviest Nuclei $^{293,294}117$ and $^{294}118$. *Phys. Rev. Lett.* **2012**, *109*, 162501. [CrossRef] [PubMed]
2. Dechargé, J.; Berger, J.F.; Dietrich, K.; Weiss, M. Superheavy and hyperheavy nuclei in the form of bubbles or semi-bubbles. *Phys. Lett. B* **1999**, *451*, 275–282. [CrossRef]
3. Weizsäcker, C.F.V. Zur Theorie der Kernmassen. *Z. Phys.* **1935**, *96*, 431–458. [CrossRef]
4. Sorlin, O.; Porquet, M.G. Nuclear magic numbers: New features far from stability. *Prog. Part. Nucl. Phys.* **2008**, *61*, 602–673. [CrossRef]
5. Otsuka, T.; Gade, A.; Sorlin, O.; Suzuki, T.; Utsuno, Y. Evolution of shell structure in exotic nuclei. *Rev. Mod. Phys.* **2020**, *92*, 015002. [CrossRef]
6. Lunney, D.; Pearson, J.M.; Thibault, C. Recent trends in the determination of nuclear masses. *Rev. Mod. Phys.* **2003**, *75*, 1021–1082. [CrossRef]
7. Yamaguchi, T.; Koura, H.; Litvinov, Y.; Wang, M. Masses of exotic nuclei. *Prog. Part. Nucl. Phys.* **2021**, *120*, 103882. [CrossRef]
8. Mumpower, M.; Surman, R.; McLaughlin, G.; Aprahamian, A. The impact of individual nuclear properties on r-process nucleosynthesis. *Prog. Part. Nucl. Phys.* **2016**, *86*, 86–126. [CrossRef]
9. Jiang, X.F.; Wu, X.H.; Zhao, P.W. Sensitivity Study of r-process Abundances to Nuclear Masses. *Astrophys. J.* **2021**, *915*, 29. [CrossRef]
10. Wu, X.H.; Zhao, P.W.; Zhang, S.Q.; Meng, J. High-precision Nuclear Chronometer for the Cosmos. *Astrophys. J.* **2022**, *941*, 152. [CrossRef]
11. Wu, X.H.; Meng, J. Supporting the CMB cosmic age from nuclear physics. *Sci. Bull.* **2023**, *68*, 539–541. [CrossRef] [PubMed]
12. Wang, M.; Huang, W.; Kondev, F.; Audi, G.; Naimi, S. The AME 2020 atomic mass evaluation (II). Tables, graphs and references. *Chin. Phys. C* **2021**, *45*, 030003. [CrossRef]
13. Pearson, J.; Nayak, R.; Goriely, S. Nuclear mass formula with Bogolyubov-enhanced shell-quenching: Application to r-process. *Phys. Lett. B* **1996**, *387*, 455–459. [CrossRef]
14. Wang, N.; Liu, M.; Wu, X.; Meng, J. Surface diffuseness correction in global mass formula. *Phys. Lett. B* **2014**, *734*, 215–219. [CrossRef]

15. Möller, P.; Sierk, A.; Ichikawa, T.; Sagawa, H. Nuclear ground-state masses and deformations: FRDM(2012). *At. Data Nucl. Data Tables* **2016**, *109–110*, 1–204. [CrossRef]
16. Koura, H.; Tachibana, T.; Uno, M.; Yamada, M. Nuclidic Mass Formula on a Spherical Basis with an Improved Even-Odd Term. *Prog. Theor. Phys.* **2005**, *113*, 305–325. [CrossRef]
17. Goriely, S.; Chamel, N.; Pearson, J.M. Skyrme-Hartree-Fock-Bogoliubov Nuclear Mass Formulas: Crossing the 0.6 MeV Accuracy Threshold with Microscopically Deduced Pairing. *Phys. Rev. Lett.* **2009**, *102*, 152503. [CrossRef]
18. Goriely, S.; Hilaire, S.; Girod, M.; Péru, S. First Gogny-Hartree-Fock-Bogoliubov Nuclear Mass Model. *Phys. Rev. Lett.* **2009**, *102*, 242501. [CrossRef]
19. Peña-Arteaga, D.; Goriely, S.; Chamel, N. Relativistic mean-field mass models. *Eur. Phys. J.* **2016**, *52*, 320. [CrossRef]
20. Xia, X.; Lim, Y.; Zhao, P.; Liang, H.; Qu, X.; Chen, Y.; Liu, H.; Zhang, L.; Zhang, S.; Kim, Y.; et al. The limits of the nuclear landscape explored by the relativistic continuum Hartree–Bogoliubov theory. *At. Data Nucl. Data Tables* **2018**, *121–122*, 1–215. [CrossRef]
21. Yang, Y.L.; Wang, Y.K.; Zhao, P.W.; Li, Z.P. Nuclear landscape in a mapped collective Hamiltonian from covariant density functional theory. *Phys. Rev. C* **2021**, *104*, 054312. [CrossRef]
22. Zhang, K.; Cheoun, M.K.; Choi, Y.B.; Chong, P.S.; Dong, J.; Dong, Z.; Du, X.; Geng, L.; Ha, E.; He, X.T.; et al. Nuclear mass table in deformed relativistic Hartree–Bogoliubov theory in continuum, I: Even–even nuclei. *At. Data Nucl. Data Tables* **2022**, *144*, 101488. [CrossRef]
23. Pan, C.; Cheoun, M.K.; Choi, Y.B.; Dong, J.; Du, X.; Fan, X.H.; Gao, W.; Geng, L.; Ha, E.; He, X.T.; et al. Deformed relativistic Hartree-Bogoliubov theory in continuum with a point-coupling functional. II. Examples of odd Nd isotopes. *Phys. Rev. C* **2022**, *106*, 014316. [CrossRef]
24. Guo, P.; Cao, X.; Chen, K.; Chen, Z.; Cheoun, M.K.; Choi, Y.B.; Lam, P.C.; Deng, W.; Dong, J.; Du, P.; et al. Nuclear mass table in deformed relativistic Hartree–Bogoliubov theory in continuum, II: Even-Z nuclei. *At. Data Nucl. Data Tables* **2024**, *158*, 101661. [CrossRef]
25. Utama, R.; Piekarewicz, J.; Prosper, H.B. Nuclear mass predictions for the crustal composition of neutron stars: A Bayesian neural network approach. *Phys. Rev. C* **2016**, *93*, 014311. [CrossRef]
26. Neufcourt, L.; Cao, Y.; Nazarewicz, W.; Olsen, E.; Viens, F. Neutron Drip Line in the Ca Region from Bayesian Model Averaging. *Phys. Rev. Lett.* **2019**, *122*, 062502. [CrossRef]
27. Wu, X.H.; Zhao, P.W. Predicting nuclear masses with the kernel ridge regression. *Phys. Rev. C* **2020**, *101*, 051301. [CrossRef]
28. Wu, X.H.; Guo, L.H.; Zhao, P.W. Nuclear masses in extended kernel ridge regression with odd-even effects. *Phys. Lett. B* **2021**, *819*, 136387. [CrossRef]
29. Guo, L.H.; Wu, X.H.; Zhao, P.W. Nuclear Mass Predictions of the Relativistic Density Functional Theory with the Kernel Ridge Regression and the Application to r-Process Simulations. *Symmetry* **2022**, *14*, 1078. [CrossRef]
30. Wu, X.H.; Lu, Y.Y.; Zhao, P.W. Multi-task learning on nuclear masses and separation energies with the kernel ridge regression. *Phys. Lett. B* **2022**, *834*, 137394. [CrossRef]
31. Du, X.K.; Guo, P.; Wu, X.H.; Zhang, S.Q. Examination of machine learning for assessing physical effects: Learning the relativistic continuum mass table with kernel ridge regression*. *Chin. Phys. C* **2023**, *47*, 074108. [CrossRef]
32. Wu, X.H.; Pan, C.; Zhang, K.Y.; Hu, J. Nuclear mass predictions of the relativistic continuum Hartree-Bogoliubov theory with the kernel ridge regression. *Phys. Rev. C* **2024**, *109*, 024310. [CrossRef]
33. Niu, Z.M.; Liang, H.Z. Nuclear mass predictions with machine learning reaching the accuracy required by *r*-process studies. *Phys. Rev. C* **2022**, *106*, L021303. [CrossRef]
34. Li, M.; Sprouse, T.M.; Meyer, B.S.; Mumpower, M.R. Atomic masses with machine learning for the astrophysical r process. *Phys. Lett. B* **2024**, *848*, 138385. [CrossRef]
35. Wu, X.H.; Zhao, P.W. Principal components of nuclear mass models. *Sci. China-Phys. Mech. Astron.* **2024**, *67*, 272011. [CrossRef]
36. Wu, X.H.; Pan, C. Nuclear mass predictions with anisotropic kernel ridge regression. *Phys. Rev. C* **2024**, *110*, 034322. [CrossRef]
37. Zhou, S.G.; Meng, J.; Ring, P.; Zhao, E.G. Neutron halo in deformed nuclei. *Phys. Rev. C* **2010**, *82*, 011301. [CrossRef]
38. Li, L.; Meng, J.; Ring, P.; Zhao, E.G.; Zhou, S.G. Deformed relativistic Hartree-Bogoliubov theory in continuum. *Phys. Rev. C* **2012**, *85*, 024312. [CrossRef]
39. Sun, X.X. Deformed two-neutron halo in ^{19}B. *Phys. Rev. C* **2021**, *103*, 054315. [CrossRef]
40. Zhang, K.; He, X.; Meng, J.; Pan, C.; Shen, C.; Wang, C.; Zhang, S. Predictive power for superheavy nuclear mass and possible stability beyond the neutron drip line in deformed relativistic Hartree-Bogoliubov theory in continuum. *Phys. Rev. C* **2021**, *104*, L021301. [CrossRef]
41. Pan, C.; Zhang, K.Y.; Chong, P.S.; Heo, C.; Ho, M.C.; Lee, J.; Li, Z.P.; Sun, W.; Tam, C.K.; Wong, S.H.; et al. Possible bound nuclei beyond the two-neutron drip line in the $50 \leq Z \leq 70$ region. *Phys. Rev. C* **2021**, *104*, 024331. [CrossRef]
42. Sun, X.X.; Zhou, S.G. Rotating deformed halo nuclei and shape decoupling effects. *Sci. Bull.* **2021**, *66*, 2072–2078. [CrossRef] [PubMed]
43. Choi, Y.B.; Lee, C.H.; Mun, M.H.; Kim, Y. Bubble nuclei with shape coexistence in even-even isotopes of Hf to Hg. *Phys. Rev. C* **2022**, *105*, 024306. [CrossRef]
44. Kim, S.; Mun, M.H.; Cheoun, M.K.; Ha, E. Shape coexistence and neutron skin thickness of Pb isotopes by the deformed relativistic Hartree-Bogoliubov theory in continuum. *Phys. Rev. C* **2022**, *105*, 034340. [CrossRef]

45. Zhang, K.; Yang, S.; An, J.; Zhang, S.; Papakonstantinou, P.; Mun, M.H.; Kim, Y.; Yan, H. Missed prediction of the neutron halo in 37Mg. *Phys. Lett. B* **2023**, *844*, 138112. [CrossRef]
46. Zhang, K.Y.; Papakonstantinou, P.; Mun, M.H.; Kim, Y.; Yan, H.; Sun, X.X. Collapse of the $N = 28$ shell closure in the newly discovered ^{39}Na nucleus and the development of deformed halos towards the neutron dripline. *Phys. Rev. C* **2023**, *107*, L041303. [CrossRef]
47. Zhang, X.Y.; Niu, Z.M.; Sun, W.; Xia, X.W. Nuclear charge radii and shape evolution of Kr and Sr isotopes with the deformed relativistic Hartree-Bogoliubov theory in continuum. *Phys. Rev. C* **2023**, *108*, 024310. [CrossRef]
48. Guo, P.; Pan, C.; Zhao, Y.C.; Du, X.K.; Zhang, S.Q. Prolate-shape dominance in atomic nuclei within the deformed relativistic Hartree-Bogoliubov theory in continuum. *Phys. Rev. C* **2023**, *108*, 014319. [CrossRef]
49. Zhang, K.Y.; Zhang, S.Q.; Meng, J. Possible neutron halo in the triaxial nucleus ^{42}Al. *Phys. Rev. C* **2023**, *108*, L041301. [CrossRef]
50. Mun, M.H.; Kim, S.; Cheoun, M.K.; So, W.; Choi, S.; Ha, E. Odd-even shape staggering and kink structure of charge radii of Hg isotopes by the deformed relativistic Hartree–Bogoliubov theory in continuum. *Phys. Lett. B* **2023**, *847*, 138298. [CrossRef]
51. Xiao, Y.; Xu, S.Z.; Zheng, R.Y.; Sun, X.X.; Geng, L.S.; Zhang, S.S. One-proton emission from 148–151Lu in the DRHBc+WKB approach. *Phys. Lett. B* **2023**, *845*, 138160. [CrossRef]
52. He, X.T.; Wu, J.W.; Zhang, K.Y.; Shen, C.W. Odd-even differences in the stability "peninsula" in the $106 \leq Z \leq 112$ region with the deformed relativistic Hartree-Bogoliubov theory in continuum. *Phys. Rev. C* **2024**, *110*, 014301. [CrossRef]
53. Lu, Q.; Zhang, K.Y.; Zhang, S.S. Triaxial shape of the one-proton emitter 149Lu. *Phys. Lett. B* **2024**, *856*, 138922. [CrossRef]
54. Pan, C.; Zhang, K.; Zhang, S. Nuclear magnetism in the deformed halo nucleus 31Ne. *Phys. Lett. B* **2024**, *855*, 138792. [CrossRef]
55. An, J.L.; Zhang, K.Y.; Lu, Q.; Zhong, S.Y.; Zhang, S.S. A unified description of the halo nucleus ^{37}Mg from microscopic structure to reaction observables. *Phys. Lett. B* **2024**, *849*, 138422. [CrossRef]
56. Zhang, K.Y.; Pan, C.; Wang, S. Examination of the evidence for a proton halo in ^{22}Al. *Phys. Rev. C* **2024**, *110*, 014320. [CrossRef]
57. DRHBc Mass Table Collaboration. Available online: http://drhbctable.jcnp.org/ (accessed on 1 October 2024).
58. Zhao, P.W.; Li, Z.P.; Yao, J.M.; Meng, J. New parametrization for the nuclear covariant energy density functional with a point-coupling interaction. *Phys. Rev. C* **2010**, *82*, 054319. [CrossRef]
59. Zhang, K.; Cheoun, M.K.; Choi, Y.B.; Chong, P.S.; Dong, J.; Geng, L.; Ha, E.; He, X.; Heo, C.; Ho, M.C.; et al. Deformed relativistic Hartree-Bogoliubov theory in continuum with a point-coupling functional: Examples of even-even Nd isotopes. *Phys. Rev. C* **2020**, *102*, 024314. [CrossRef]
60. Kucharek, H.; Ring, P. Relativistic field theory of superfluidity in nuclei. *Z. Phys. A* **1991**, *339*, 23–35. [CrossRef]
61. Zhou, S.G.; Meng, J.; Ring, P. Spherical relativistic Hartree theory in a Woods-Saxon basis. *Phys. Rev. C* **2003**, *68*, 034323. [CrossRef]
62. Ring, P.; Schuck, P. *The Nuclear Many-Body Problem*; Springer: New York, NY, USA, 1980.
63. Pan, C.; Zhang, K.; Zhang, S. Multipole expansion of densities in the deformed relativistic Hartree-Bogoliubov theory in continuum. *Int. J. Mod. Phys. E* **2019**, *28*, 1950082. [CrossRef]
64. Chatt, J. Recommendations for the naming of elements of atomic numbers greater than 100. *Pure Appl. Chem.* **1979**, *51*, 381–384.
65. Wang, S.; Guo, P.; Cong, P. Determining the ground state for superheavy nuclei from the deformed relativistic Hartree-Bogoliubov theory in continuum. *Particles* 11. *under review*.

particles

MDPI

Article

Masses and Quadrupole Deformations of Even-Z Nuclei Within a Triaxial Relativistic Hartree–Bogoliubov Model

Qin Zhou and Zhipan Li *

School of Physical Science and Technology, Southwest University, Chongqing 400715, China; zqxhxxd@163.com
* Correspondence: zpliphy@swu.edu.cn

Abstract: This study investigates the masses and quadrupole deformations of even-Z nuclei within the range $8 \leqslant Z \leqslant 104$ using the triaxial relativistic Hartree–Bogoliubov model (TRHB) with the PC-PK1 density functional. For odd-mass nuclei, the global minima were determined using the automatic blocking method and their dynamical correlation energies (DCEs) were approximated using the average values of neighboring even–even nuclei calculated from a microscopic, five-dimensional, collective Hamiltonian (5DCH). The mean-field results underestimate the binding energies of most open-shell nuclei, with an initial root–mean–square (rms) deviation of 2.56 MeV for 1223 even-Z nuclei. Incorporating DCEs significantly reduces this deviation to 1.36 MeV. Additionally, the descriptions of two-neutron and one-neutron separation energies are improved, with rms deviations decreasing to 0.75 MeV and 0.65 MeV, respectively. Further refinement through accounting for odd–even differences in DCEs reduces the rms deviations for binding energies and one-neutron separation energies to 1.30 MeV and 0.63 MeV, respectively. Regarding the quadrupole deformations, TRHB calculations reveal spherical shapes near shell and subshell closures, well-deformed shapes at the mid-shell, and rapid shape transitions in medium- and heavy-mass regions. Oblate shapes dominate in regions $(Z, N) \sim (14, 14)$, $(34, 36)$, and $(40, 60)$, and the neutron-deficient Pb region, with notable odd–even shape staggering attributed to the blocking effect of the odd nucleon. Triaxial shapes are favored in the mass regions $(Z, N) \sim (60, 76)$ and $(76, 116)$.

Keywords: nuclear mass; quadrupole deformation; triaxial relativistic Hartree–Bogoliubov; blocking effect; five-dimensional collective Hamiltonian

1. Introduction

Nuclear mass (or binding energy) is a fundamental property of atomic nuclei. To date, more than 3300 nuclides have been identified or synthesized in laboratories [1], with approximately 300 occurring naturally. Using advanced experimental techniques such as time-of-flight measurements, storage ring mass spectrometry, and ion trap methods [2,3], researchers have accurately determined the masses of more than 2500 nuclides [4].

Theoretical models predict that there are approximately 7000 to 12,000 bound nuclei [5,6], with the location of the neutron drip line for $Z > 11$ remaining highly uncertain. The search for the limits of nuclear binding is deeply intertwined with the need to understand the origin of elements [7]. In particular, the astrophysical rapid neutron-capture process, which is responsible for the formation of many heavy elements, critically depends on the properties of nuclei near the neutron drip line. These properties, however, will remain experimentally inaccessible in the foreseeable future, making theoretical predictions across the nuclear landscape indispensable. Over the past few decades, significant progress has been made using macroscopic–microscopic models [8–11], achieving remarkable accuracy with root–mean–square deviations of approximately 300 keV [12].

Nuclear density functional theory (DFT), based on effective nucleon–nucleon interactions, is a highly promising microscopic approach in the field of nuclear physics. It is built upon a universal energy density functional and, with just a limited number of parameters, it is able to describe not only the nuclear binding energies but also other crucial quantities relevant to simulating the nucleosynthesis process, such as β-decay rates and fission rates, in a unified and comprehensive manner [13].

A number of Hartree–Fock–Bogoliubov (HFB) mass-table-type calculations have been performed based on the non-relativistic Skyrme [5,14–18] and Gogny [19–21] density functionals. The Skyrme mass model HFB-27* [18] determines its 24 parameters in functionals and corrections by fitting 2353 nuclear masses, achieving an accuracy of 0.512 MeV. The Gogny mass model D1M [20] incorporates the beyond-mean-field dynamical correlation energies (DCEs) and fits its 14 parameters to 2149 measured masses, resulting in a final rms deviation of 0.798 MeV.

The relativistic DFT was proven to be a powerful theory in nuclear physics and has been successfully employed to describe a variety of nuclear phenomena [22–30]. Significant progress in mass description has also been achieved based on relativistic DFT. In 2015, we performed a systematic calculation for 575 even–even nuclei using the relativistic mean-field plus Bardeen–Cooper–Schrieffer (RMF+BCS) framework with the PC-PK1 functional [31], and the beyond-mean-field DCEs were taken into account using a microscopically mapped, five-dimensional, collective Hamiltonian (5DCH) without additional free parameters. The root–mean–square (rms) deviation of nuclear masses was significantly reduced to 1.14 MeV [32]. In 2021, we extended our previous work to the nuclear landscape of even–even nuclei using an advanced triaxial relativistic Hartree–Bogoliubov (TRHB) framework [33]. It was emphasized that the nuclear landscape is considerably extended using the PC-PK1 functional compared to previous results obtained with the relativistic density functionals DD-PC1 [34] and TMA [35]. Moreover, the calculations reproduced the measured mass regions for the shape and the coexistence of multiple shapes and predict new regions [36].

Our final goal is to establish a global calculation for the nuclear ground state and low-lying excitation properties across the entire nuclear landscape using the TRHB with the PC-PK1 functional. As mentioned above, for the first step, we fulfilled this aim for even–even nuclei [33,36]. Here, we will conduct a study of the properties of odd-mass nuclei, focusing on the masses and quadrupole deformations of the even-Z odd-N nuclei whose masses have been measured. The key aspect is the appropriate handling of the blocking of odd nucleon and the beyond-mean-field DCEs. The blocking of odd nucleon was introduced in the TRHB model and successfully applied to numerous studies, such as the Yrast band of ^{109}Ag [37], magnetic rotations in ^{198}Pb and ^{199}Pb [38], chirality in ^{106}Ag [39], and nuclear $\beta\beta$ decay [40]. As the variation in DCEs is gentle [33,41], in this work, we will approximately calculate the DCEs of odd-mass nuclei through the interpolation of those of neighboring even–even nuclei, and further phenomenologically take into account the odd–even differences in DCEs.

In Section 2, we introduce the Relativistic Hartree–Bogoliubov (RHB) theory and the blocking effect for odd-mass nuclei. Section 3 presents a systematic analysis of the binding energies, quadrupole deformations, and dynamical correlation energies (DCEs) of 1223 even-Z nuclei. Finally, the conclusions and future perspectives are summarized.

2. Theoretical Framework and Numerical Details

The RHB theory provides a unified and self-consistent treatment of mean fields and pairing correlations [42–45]. The RHB equation is as follows:

$$\begin{pmatrix} h_D - \lambda_\tau & \Delta \\ -\Delta^* & -h_D^* + \lambda_\tau \end{pmatrix} \begin{pmatrix} U_k \\ V_k \end{pmatrix} = E_k \begin{pmatrix} U_k \\ V_k \end{pmatrix}, \tag{1}$$

where h_D is the Dirac Hamiltonian, Δ is the pairing field, λ_τ is the Fermi energy for neutron or proton ($\tau = n, p$), E_k is the quasiparticle energy, and U_k and V_k are the quasiparticle wave functions.

The Dirac Hamiltonian in the coordinate space is

$$h_D(\boldsymbol{r}) = \boldsymbol{\alpha} \cdot \boldsymbol{p} + V(\boldsymbol{r}) + \beta[M + S(\boldsymbol{r})], \tag{2}$$

with the scalar and vector potentials

$$\begin{aligned} S(\boldsymbol{r}) &= \alpha_S \rho_S + \beta_S \rho_S^2 + \gamma_S \rho_S^3 + \delta_S \Delta \rho_S, \\ V(\boldsymbol{r}) &= \alpha_V \rho_V + \gamma_V \rho_V^3 + \delta_V \Delta \rho_V + eA^0 + \alpha_{TV} \tau_3 \rho_3 + \delta_{TV} \tau_3 \Delta \rho_3 \end{aligned} \tag{3}$$

constructed using the following densities:

$$\begin{aligned} \rho_S(\boldsymbol{r}) &= \sum_{k>0} V_k^\dagger(\boldsymbol{r}) \gamma_0 V_k(\boldsymbol{r}), \\ \rho_V(\boldsymbol{r}) &= \sum_{k>0} V_k^\dagger(\boldsymbol{r}) V_k(\boldsymbol{r}), \\ \rho_3(\boldsymbol{r}) &= \sum_{k>0} V_k^\dagger(\boldsymbol{r}) \tau_3 V_k(\boldsymbol{r}). \end{aligned} \tag{4}$$

According to the no-sea approximation, the summations in the above equations were performed over the quasiparticle states with positive energies in the Fermi sea. In Equations (3), we use PC-PK1 parametrization [31] for the parameters $\alpha_S, \beta_S, \cdots$.

The pairing potential is as follows:

$$\Delta(\mathbf{r}_1, \mathbf{r}_2) = V^{pp}(\mathbf{r}_1, \mathbf{r}_2) \kappa(\mathbf{r}_1, \mathbf{r}_2), \tag{5}$$

where, for simplicity, the spin and isospin degrees of freedom are not shown, and $\kappa = V^* U^T$ is the pairing tensor [46]. In principle, the same relativistic interaction used in Equations (2) and (3) can be applied to the pairing channel. In 1991, Kucharek and Ring [43] derived the Dirac–Hartree–Fock–Bogoliubov equations using a unified relativistic interaction within a quantum field theory framework. However, their analysis of pairing correlations in symmetric nuclear matter yielded excessively large pairing gaps due to the strongly repulsive nature of the relativistic nuclear interaction. Subsequent studies demonstrated that adopting the finite-range Gogny interaction or a separable pairing force [47] for the pairing channel produces more realistic pairing gaps. Therefore, in this work, we adopted the latter approach, employing the separable pairing force for V^{pp}:

$$V^{pp}(\mathbf{r}_1, \mathbf{r}_2, \mathbf{r}_1', \mathbf{r}_2') = G\delta(\mathbf{R} - \mathbf{R}') P(\mathbf{r}) P(\mathbf{r}') \frac{1}{2}(1 - P^\sigma). \tag{6}$$

The selected pairing interaction fails to maintain covariance. However, since a fully covariant treatment of paring dynamics in RMF theory is still an open question, we followed the widely practiced approach and used Equation (6) in the nuclear rest frame. Here,

$\mathbf{R} = \frac{1}{2}(\mathbf{r}_1 + \mathbf{r}_2)$ and $\mathbf{r} = \mathbf{r}_1 - \mathbf{r}_2$ denote the center of mass and the relative coordinates, respectively, and $P(\mathbf{r})$ has a Gaussian expression

$$P(\mathbf{r}) = \frac{1}{(4\pi a^2)^{3/2}} e^{-r^2/4a^2}. \tag{7}$$

The projector $\frac{1}{2}(1 - P^{\sigma})$ allows only the states with a total spin $S = 0$. The two parameters G and a were determined in Ref. [47] through fitting to the density dependence of pairing gaps at the Fermi surface for nuclear matter obtained with the Gogny forces.

For an odd-mass nucleus, the blocking effect of the unpaired nucleon needs to be considered [46]. The ground state of a system with an unpaired particle can be described using a one-quasiparticle state:

$$|\Phi_1\rangle = \alpha^{\dagger}_{k_b}|\Phi\rangle = \alpha^{\dagger}_{k_b}\prod_k \alpha_k|0\rangle, \tag{8}$$

where $\alpha^{\dagger}_{k_b}$ corresponds to the quasiparticle state that is properly blocked. In other words, the one-quasiparticle state $|\Phi_1\rangle$ is the vacuum with respect to the set of quasiparticle operators $(\alpha_1, \ldots, \alpha^{\dagger}_{k_b}, \ldots, \alpha_N)$. That is, the blocking effect can be realized through the exchange of $\alpha_{k_b} \leftrightarrow \alpha^{\dagger}_{k_b}$. According to Equation (1), this exchange corresponds to the exchange of the columns $(V^*_{k_b}, U^*_{k_b}) \leftrightarrow (U_{k_b}, V_{k_b})$ and that of the energy $E_{k_b} \leftrightarrow -E_{k_b}$. In principle, the blocked orbital k_b breaks the time reversal symmetry and the currents appear. However, in general, the contribution of the currents is very weak for the bulk properties of odd-mass nuclei [48], and thus the currents were neglected in the present calculation. To determine the ground state of an odd-mass nucleus, one needs to find the correct deformation minimum with the correct blocking orbital. Here, we blocked the lowest quasiparticle orbital in each iteration of TRHB equation, which is referred to as "automatic blocking". Moreover, for the shape-coexisting nuclei or soft nuclei, one has to perform automatic blocking calculations staring from various initial states with different deformations.

The TRHB equation is solved through expanding the quasiparticle wavefunctions in terms of a three-dimensional harmonic oscillator basis in Cartesian coordinates [49]. For nuclei with $Z < 20$, $20 \leq Z < 82$, and $82 \leq Z \leq 104$, the harmonic oscillator basis contains 12, 14, and 16 major shells, respectively, and these were examined to determine their ability to provide converged results.

3. Results and Discussion

Figure 1 illustrates the differences between the experimental binding energies [4] and the theoretical ones calculated by the TRHB using the PC-PK1 functional for 1223 even-Z nuclei. The mean-field results for the nuclei around the neutron shell closure ($N = 8, 20, 50, 82$) and proton shell closure ($Z = 20$) agree well with the data. However, the binding energies were underestimated for most open-shell nuclei, and large discrepancies (>3 MeV) were found in the transitional regions (c.f. Figure 2). The root–mean–square (rms) deviation from the experimental nuclear binding energies is 2.56 MeV.

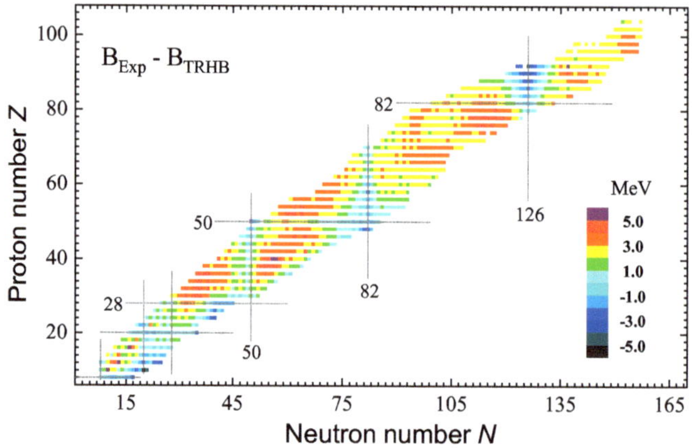

Figure 1. Differences between the experimental binding energies [4] and the theoretical ones calculated by the TRHB with the PC-PK1 functional for 1223 even-Z nuclei. Considering the non-convergency of PC-PK1 for ^{14}O, $^{23-26}$Mg, ^{26,27}Si, ^{33,34}S, and ^{34}Ar, the PCF-PK1 functional [50] was adopted for these nuclei.

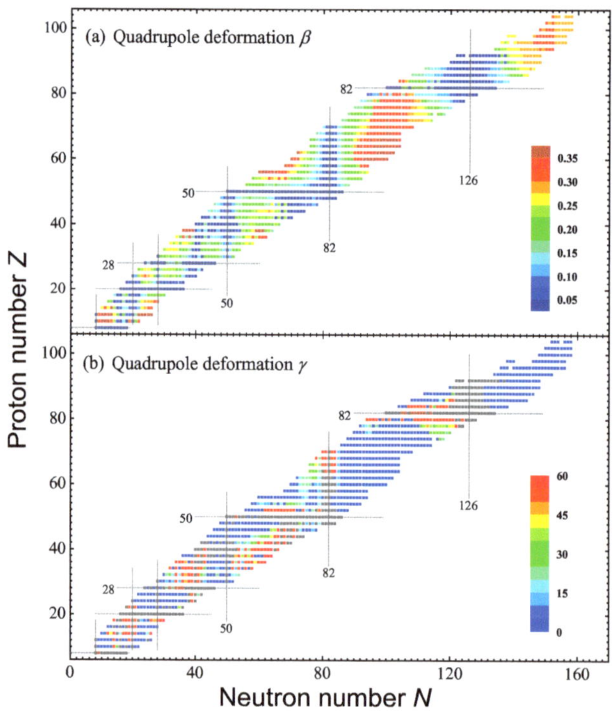

Figure 2. Quadrupole deformations β (panel (**a**)) and γ (panel (**b**)) of the mean-field ground state, calculated using TRHB with the PC-PK1 functional, are shown for 1223 even-Z nuclei. The gray color in panel (**b**) denotes the spherical nuclei, where γ is meaningless.

Figure 2 displays the quadrupole deformations β and γ of the mean-field ground state, calculated using TRHB with PC-PK1 functional. Here, β represents the quadrupole deformation, which is distinct from the 4×4 Dirac matrix β in Equation (2). In general, the nuclei close to the shells and subshell $(N, Z = 8, 20, 28, 40, 50, 82, 126)$ are spherical, while the mid-shell nuclei are well-deformed. Between these areas, rapid shape transitions or even shape phase transitions can be observed in the medium- and heavy-mass regions. Shape phase transitions are characterized by abrupt changes in the ground-state nuclear shape. In panel (b), most of the deformed nuclei are prolate and oblate shapes are favored in the regions $(Z,N) \sim (14,14),(34,36)$, and $(40,60)$, as well as in neutron-deficient Pb region, which are well-known to contain coexisting shapes [51,52]. Moreover, in these regions, the quadrupole deformations in the odd nuclei are quite different from those of the neighboring even nuclei, presenting with an odd–even staggered shape. This is due the subtle balance between the coexisting shapes with different blocking configurations occupied by the odd nucleon [53]. These results are consistent with those of DRHBc mass table [41] and HFB calculations using Gogny D1S [21]. In panel (b), it is also interesting to observe that triaxial shapes are favored in the mass regions $(Z,N) \sim (60,76)$ and $(76,116)$.

The TRHB results were obtained based on the static mean-field approximation, which is characterized through the breaking of symmetries in the underlying Hamiltonian, including translational, rotational, and particle number symmetries. This leads to the inclusion of static correlations such as deformations and pairing. However, to incorporate the dynamical correlations, one has to move beyond the mean-field approximation by restoring the broken symmetries and mixing the configurations of symmetry-breaking product states. For even–even nuclei, we utilized a TRHB-based, five-dimensional collective Hamiltonian (5DCH) to take into account the quadrupole dynamical correlations. The resulting ground-state energy of the 0_1^+ state (laboratory frame) is derived without phenomenological adjustments, providing a direct comparison with the experimental data [33,36]. The dynamical correlation energy (DCE) can be defined as the energy difference between the mean-field global minimum in the intrinsic frame and the ground state with good angular momentum in the laboratory frame, representing the energy correction from the dynamical collective correlations. This aligns with the standard many-body techniques for symmetry restoration.

In Figure 3a, we take even–even Th isotopes as examples to illustrate the DCEs calculated from 5DCH based on TRHB with a PC-PK1 functional. These vary in a relatively mild manner with the neutron number, except for the nuclei close to the neutron shells. For the odd mass nuclei, recently, rotational DCEs were extracted using the state-of-the-art angular momentum projection method, and it was found that they change smoothly from a light- to a heavy-mass region [54]. They also proved the validity of the cranking approximation for the odd-mass nuclei, which is used to construct a DRHBc mass table for the even-Z nuclei. Figure 3b displays the odd–even differences $\delta_{\rm DCE}(N) = (-1)^N(2 \times \mathrm{DCE}(N) - \mathrm{DCE}(N-1) - \mathrm{DCE}(N+1))/2$ for the rotational DCEs, which were taken from the DRHBc mass table [41]. These are basically oscillated around 0.2 MeV. For nuclei in the gaps in Figure 3b, the DCE was set to zero, and the corresponding odd–even differences $\delta_{\rm DCE}(N)$ were consequently assigned a value of zero.

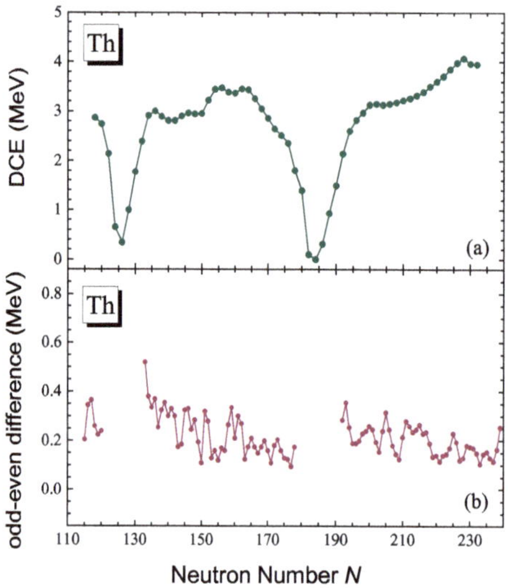

Figure 3. (**a**) DCEs of the even–even Th isotopic chain calculated using 5DCH based on TRHB as a function of neutron number N. (**b**) Odd–even difference $\delta_{\text{DCE}}(N) = (-1)^N(2 \times \text{DCE}(N) - \text{DCE}(N-1) - \text{DCE}(N+1))/2$ for the rotational DCEs of Th isotopes. The rotational DCEs are taken from the DRHBc mass table [41].

The above findings enable us to approximate the DCEs of odd-mass nuclei by averaging those of their neighboring even–even nuclei. In Figure 4 and Table 1, we present the differences between the experimental binding energies [4] and the theoretical values incorporating DCEs. Notably, the theoretical results exhibit excellent agreement with the experimental data, with the energy deviations for most nuclei falling within ± 1 MeV. Larger deviations (>3 MeV) were primarily observed in light nuclei and the $N \sim 126$ mass region. For the 1223 even-Z nuclei, the rms deviation from the experimental binding energies was significantly reduced to 1.36 MeV. Furthermore, in Figure 5 and Table 1, the description for both the two-neutron and one-neutron separation energies was optimized by considering the DCEs, and their rms deviations were reduced from 0.98 and 0.75 MeV to 0.75 and 0.65 MeV, respectively. Notably, this accurate description of nuclear binding energies was achieved using a global density functional, without introducing any additional phenomenological parameters.

Finally, we considered the odd–even differences for DCEs through phenomenologically subtracting a constant from those of odd-mass nuclei. For odd-mass nuclei with an initial DCE = 0 or those whose DCEs would become negative after subtraction (i.e., DCE < 0), the DCEs were clamped at zero to ensure physical consistency. When the constant was chosen as 0.20 MeV, we obtained the best description for the binding energies and S_n, whose rms deviations were further reduced to 1.30 and 0.63 MeV, respectively (see the last row of Table 1). This serveed as a valuable reference for future refinements of odd-A nuclei.

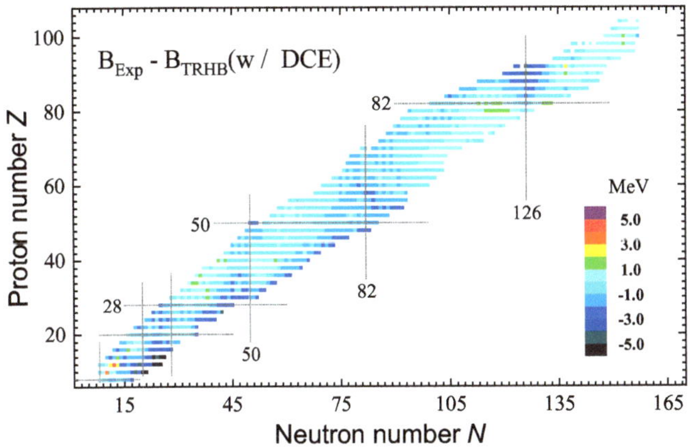

Figure 4. Differences between the experimental binding energies [4] and the theoretical ones calculated using the TRHB with DCEs for 1223 even-Z nuclei.

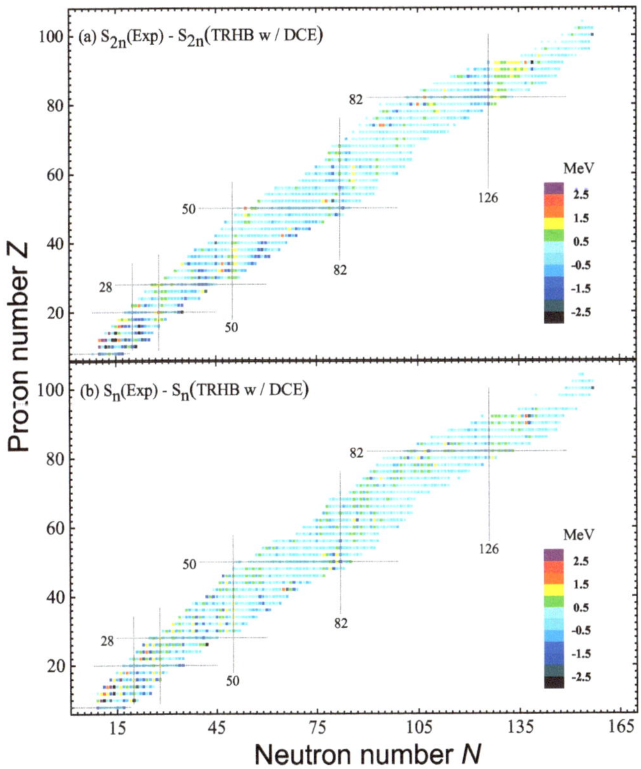

Figure 5. Differences between the experimental separation energies [4] and the theoretical ones calculated using the TRHB with DCEs for 1223 even-Z nuclei. The two-neutron separation energies S_{2n} are shown in panel (**a**), while the one-neutron separation energies S_n are shown in panel (**b**).

Table 1. The rms deviations for the binding energies, two-neutron separation energies S_{2n}, and one-neutron separation energies S_n in the TRHB calculations without (first row) and with (second row) DCEs, in comparison with the AME2020 data [4]. The dataset comprises 1223 nuclei from AME2020 data [4], including 606 even-Z odd-N nuclei and 617 neighboring even–even nuclei. The DCE* in the last row means that all the DCEs of 617 odd-mass nuclei were subtracted with a constant 0.20 MeV. All the values in the table are in the unit of MeV.

DFTs	σ_B	$\sigma_{S_{2n}}$	σ_{S_n}	Date Number
TRHB w/o DCE	2.56	0.98	0.75	1223
TRHB w DCE	1.36	0.75	0.65	1223
TRHB w DCE*	1.30	0.75	0.63	1223

4. Summary and Outlook

In summary, the masses and quadrupole deformations of even-Z nuclei within the range $8 \leqslant Z \leqslant 104$ were calculated using the triaxial relativistic Hartree–Bogoliubov model with the PC-PK1 density functional. For odd-mass nuclei, the global minima were determined via the automatic blocking method, while their dynamical correlation energies (DCEs) were approximated using the average values of adjacent even–even nuclei, derived from a microscopic five-dimensional collective Hamiltonian (5DCH) [33].

The mean-field results underestimate the nuclear binding energies for most open-shell nuclei, and the root–mean–square (rms) deviation is 2.56 MeV for the total 1223 even-Z nuclei. After incorporating dynamical correlation energies (DCEs), the rms deviation is significantly reduced to 1.36 MeV. Additionally, the description of two-neutron and one-neutron separation energies improves, with rms deviations decreasing from 0.98 MeV and 0.75 MeV to 0.75 MeV and 0.65 MeV, respectively. Further refinement is achieved by accounting for odd–even differences in DCEs through a phenomenological adjustment, where a constant 0.20 MeV is subtracted from the DCEs of odd-mass nuclei. This reduces the rms deviations for binding energies and one-neutron separation energies to 1.30 MeV and 0.63 MeV, respectively.

Regarding quadrupole deformations, nuclei near-shell and subshell closures have spherical shapes, while those at the mid-shell are very deformed. In the medium- and heavy-mass regions, rapid shape transitions—or even shape-phase transitions—are observed. Triaxial shapes are predicted in the mass regions $(Z, N) \sim (60, 76)$ and $(76, 116)$. Oblate shapes are favored in the regions $(Z, N) \sim (14, 14), (34, 36)$, and $(40, 60)$, as well as in the neutron-deficient Pb region, where we also can also observe that the quadrupole deformations of odd nuclei differ significantly from those of the neighboring even nuclei, manifesting as an odd–even shape staggering. This phenomenon arises due to the delicate balance between coexisting shapes, influenced by the blocking configurations of the odd nucleon [53].

In the future, we will continue to investigate the covariance issues in the pairing interaction. The microscopic 5DCH framework can be extended to explicitly account for the contributions of odd nucleons. This would enable precise calculations of dynamical correlation energies (DCEs) for odd-mass and odd–odd nuclei, as well as the construction of a comprehensive nuclear landscape based on the PC-PK1 density functional or its advanced counterpart, PCF-PK1 [50].

Author Contributions: Writing—original draft preparation, Q.Z.; writing—review and editing, Z.L.; All authors have read and agreed to the published version of the manuscript.

Funding: This work was partly supported by the National Natural Science Foundation of China (Grant No. 12375126) and the Fundamental Research Funds for the Central Universities.

Data Availability Statement: The dataset can be accessed upon request to the corresponding author.

Acknowledgments: Q. Zhou gratefully acknowledges the fruitful discussions and valuable assistance from P. W. Zhao, Y. K. Wang, Y. L. Yang, and T. Qu at Peking University, as well as M. H. Zhou at Southwest University. The helpful discussions with members of the DRHBc Mass Table Collaboration are highly appreciated.

Conflicts of Interest: The authors declare no conflicts of interest.

References

1. Thoennessen, M. *The Discovery of Isotopes*; Springer: Cham, Switzerland, 2016.
2. Wang, M. Nuclear Mass Measurement and Evaluation. *Nucl. Phys. Rev.* **2017**, *34*, 380–386.
3. Wang, M.; Zhang, Y.H.; Zhou, X.H. Nuclear mass measurements. *Sci. Sin. Phys. Mech. Astron.* **2020**, *50*, 052006.
4. Wang, M.; Huang, W.J.; Kondev, F.G.; Audi, G.; Naimi, S. The AME 2020 atomic mass evaluation (II). Tables, graphs and references. *Chin. Phys. C* **2021**, *45*, 030003. [CrossRef]
5. Erler, J.; Birge, N.; Kortelainen, M.; Nazarewicz, W.; Olsen, E.; Perhac, A.M.; Stoitsov, M. The limits of the nuclear landscape. *Nature* **2012**, *486*, 509–512. [CrossRef]
6. Xia, X.W.; Lim, Y.; Zhao, P.W.; Liang, H.Z.; Qu, X.Y.; Chen, Y.; Liu, H.; Zhang, L.F.; Zhang, S.Q.; Kim, Y.; et al. The limits of the nuclear landscape explored by the relativistic continuum Hartree—Bogoliubov theory. *At. Data Nucl. Data Tables* **2018**, *121–122*, 1–215. [CrossRef]
7. Burbidge, E.M.; Burbidge, G.R.; Fowler, W.A.; Hoyle, F. Synthesis of the elements in stars. *Rev. Mod. Phys.* **1957**, *29*, 547. [CrossRef]
8. Möller, P.; Myers, W.; Swiatecki, W.; Treiner, J. Nuclear mass formula with a finite-range droplet model and a folded-Yukawa single-particle potential. *At. Data Nucl. Data Tables* **1988**, *39*, 225–233. [CrossRef]
9. Moller, P.; Nix, J.; Myers, W.; Swiatecki, W. Nuclear Ground-State Masses and Deformations. *At. Data Nucl. Data Tables* **1995**, *59*, 185–381. [CrossRef]
10. Möller, P.; Sierk, A.; Ichikawa, T.; Sagawa, H. Nuclear ground-state masses and deformations: FRDM(2012). *At. Data Nucl. Data Tables* **2016**, *109–110*, 1–204. [CrossRef]
11. Haustein, P.E. An overview of the 1986–1987 atomic mass predictions. *At. Data Nucl. Data Tables* **1988**, *39*, 185–200. [CrossRef]
12. Wang, N.; Liu, M.; Wu, X.Z.; Meng, J. Surface diffuseness correction in global mass formula. *Phys. Lett. B* **2014**, *734*, 215–219. [CrossRef]
13. Bender, M.; Heenen, P.H.; Reinhard, P.G. Self-consistent mean-field models for nuclear structure. *Rev. Mod. Phys.* **2003**, *75*, 121. [CrossRef]
14. Samyn, M.; Goriely, S.; Heenen, P.H.; Pearson, J.; Tondeur, F. A hartree–fock–bogoliubov mass formula. *Nucl. Phys. A* **2002**, *700*, 142–156. [CrossRef]
15. Stoitsov, M.; Dobaczewski, J.; Nazarewicz, W.; Pittel, S.; Dean, D. Systematic study of deformed nuclei at the drip lines and beyond. *Phys. Rev. C—Nucl. Phys.* **2003**, *68*, 054312. [CrossRef]
16. Goriely, S.; Chamel, N.; Pearson, J. Skyrme-Hartree-Fock-Bogoliubov Nuclear Mass Formulas: Crossing the 0.6 MeV Accuracy Threshold with Microscopically Deduced Pairing. *Phys. Rev. Lett.* **2009**, *102*, 152503. [CrossRef]
17. Goriely, S.; Chamel, N.; Pearson, J. Further explorations of Skyrme-Hartree-Fock-Bogoliubov mass formulas. XIII. The 2012 atomic mass evaluation and the symmetry coefficient. *Phys. Rev. C—Nucl. Phys.* **2013**, *88*, 024308. [CrossRef]
18. Goriely, S.; Chamel, N.; Pearson, J. Hartree-Fock-Bogoliubov nuclear mass model with 0.50 MeV accuracy based on standard forms of Skyrme and pairing functionals. *Phys. Rev. C—Nucl. Phys.* **2013**, *88*, 061302. [CrossRef]
19. Hilaire, S.; Girod, M. Large-scale mean-field calculations from proton to neutron drip lines using the D1S Gogny force. *Eur. Phys. J. A* **2007**, *33*, 237–241. [CrossRef]
20. Goriely, S.; Hilaire, S.; Girod, M.; Peru, S. The Gogny-Hartree-Fock-Bogoliubov nuclear-mass model. *Phys. Rev. Lett.* **2009**, *102*, 242501. [CrossRef]
21. Delaroche, J.P.; Girod, M.; Libert, J.; Goutte, H.; Hilaire, S.; Péru, S.; Pillet, N.; Bertsch, G. Structure of even-even nuclei using a mapped collective Hamiltonian and the D1S Gogny interaction. *Phys. Rev. C—Nucl. Phys.* **2010**, *81*, 014303. [CrossRef]
22. Ring, P. Relativistic mean field theory in finite nuclei. *Prog. Part. Nucl. Phys.* **1996**, *37*, 193–263. [CrossRef]
23. Vretenar, D.; Afanasjev, A.; Lalazissis, G.; Ring, P. Relativistic Hartree–Bogoliubov theory: Static and dynamic aspects of exotic nuclear structure. *Phys. Rep.* **2005**, *409*, 101–259. [CrossRef]
24. Meng, J.; Toki, H.; Zhou, S.G.; Zhang, S.Q.; Long, W.H.; Geng, L.S. Relativistic continuum Hartree Bogoliubov theory for ground-state properties of exotic nuclei. *Prog. Part. Nucl. Phys.* **2006**, *57*, 470–563. [CrossRef]
25. Nikšić, T.; Vretenar, D.; Ring, P. Relativistic nuclear energy density functionals: Mean-field and beyond. *Prog. Part. Nucl. Phys.* **2011**, *66*, 519–548. [CrossRef]
26. Meng, J.; Peng, J.; Zhang, S.Q.; Zhao, P.W. Progress on tilted axis cranking covariant density functional theory for nuclear magnetic and antimagnetic rotation. *Front. Phys.* **2013**, *8*, 55–79. [CrossRef]

27. Meng, J.; Zhou, S.G. Halos in medium-heavy and heavy nuclei with covariant density functional theory in continuum. *J. Phys. G Nucl. Part. Phys.* **2015**, *42*, 093101. [CrossRef]
28. Zhou, S.G. Multidimensionally constrained covariant density functional theories–nuclear shapes and potential energy surfaces. *Phys. Scr.* **2016**, *91*, 063008. [CrossRef]
29. Meng, J. *Relativistic Density Functional for Nuclear Structure*; World Scientific: Singapore, 2016; Volume 10.
30. Shen, S.H.; Liang, H.Z.; Long, W.H.; Meng, J.; Ring, P. Towards an ab initio covariant density functional theory for nuclear structure. *Prog. Part. Nucl. Phys.* **2019**, *109*, 103713. [CrossRef]
31. Zhao, P.W.; Li, Z.P.; Yao, J.M.; Meng, J. New parametrization for the nuclear covariant energy density functional with a point-coupling interaction. *Phys. Rev. C* **2010**, *82*, 054319. [CrossRef]
32. Lu, K.Q.; Li, Z.X.; Li, Z.P.; Yao, J.M.; Meng, J. Global study of beyond-mean-field correlation energies in covariant energy density functional theory using a collective Hamiltonian method. *Phys. Rev. C* **2015**, *91*, 027304. [CrossRef]
33. Yang, Y.L.; Wang, Y.K.; Zhao, P.W.; Li, Z.P. Nuclear landscape in a mapped collective Hamiltonian from covariant density functional theory. *Phys. Rev. C* **2021**, *104*, 054312. [CrossRef]
34. Agbemava, S.; Afanasjev, A.; Ray, D.; Ring, P. Global performance of covariant energy density functionals: Ground state observables of even-even nuclei and the estimate of theoretical uncertainties. *Phys. Rev. C* **2014**, *89*, 054320. [CrossRef]
35. Geng, L.S.; Toki, H.; Meng, J. Masses, Deformations and Charge Radii—Nuclear Ground-State Properties in the Relativistic Mean Field Model. *Prog. Theor. Phys.* **2005**, *113*, 785–800. [CrossRef]
36. Yang, Y.L.; Zhao, P.W.; Li, Z.P. Shape and multiple shape coexistence of nuclei within covariant density functional theory. *Phys. Rev. C* **2023**, *107*, 024308. [CrossRef]
37. Wang, Y.K. Yrast band of ^{109}Ag described by tilted axis cranking covariant density functional theory with a separable pairing force. *Phys. Rev. C* **2017**, *96*, 054324. [CrossRef]
38. Wang, Y.K. Magnetic rotations in ^{198}Pb and ^{199}Pb within covariant density functional theory with pairing correlations. *Phys. Rev. C* **2018**, *97*, 064321. [CrossRef]
39. Zhao, P.W.; Wang, Y.K.; Chen, Q.B. Microscopic resolution of the nuclear chiral conundrum with crossing twin bands in ^{106}Ag. *Phys. Rev. C* **2019**, *99*, 054319. [CrossRef]
40. Wang, Y.K.; Zhao, P.W.; Meng, J. Relativistic configuration-interaction density functional theory: Nonaxial effects on nuclear $\beta\beta$ decay. *Sci. Bull.* **2024**, *69*, 2017–2020. [CrossRef] [PubMed]
41. Guo, P.; Cao, X.J.; Chen, K.M.; Chen, Z.H.; Cheoun, M.K.; Choi, Y.B.; Lam, P.C.; Deng, W.M.; Dong, J.M.; Du, P.X.; et al. Nuclear mass table in deformed relativistic Hartree–Bogoliubov theory in continuum, II: Even-Z nuclei. *At. Data Nucl. Data Tables* **2024**, *158*, 101661. [CrossRef]
42. Meng, J. Relativistic continuum Hartree-Bogoliubov theory with both zero range and finite range Gogny force and their application. *Nucl. Phys. A* **1998**, *635*, 3–42. [CrossRef]
43. Kucharek, H.; Ring, P. Relativistic field theory of superfluidity in nuclei. *Z. Für Phys. A Hadron. Nucl.* **1991**, *339*, 23–35. [CrossRef]
44. Gonzalez-Llarena, T.; Egido, J.; Lalazissis, G.; Ring, P. Relativistic Hartree-Bogoliubov calculations with finite range pairing forces. *Phys. Lett. B* **1996**, *379*, 13–19. [CrossRef]
45. Serra, M.; Ring, P. Relativistic Hartree-Bogoliubov theory for finite nuclei. *Phys. Rev. C* **2002**, *65*, 064324. [CrossRef]
46. Ring, P.; Schuck, P. *The Many Body Nuclear Problem*; Springer: Berlin/Heidelberg, Germany, 1980.
47. Tian, Y.; Ma, Z.Y.; Ring, P. A finite range pairing force for density functional theory in superfluid nuclei. *Phys. Lett. B* **2009**, *676*, 44–50. [CrossRef]
48. Pan, C.; Zhang, K.Y.; Zhang, S.Q. Nuclear magnetism in the deformed halo nucleus ^{31}Ne. *Phys. Lett. B* **2024**, *855*, 138792. [CrossRef]
49. Nikšić, T.; Paar, N.; Vretenar, D.; Ring, P. DIRHB—A relativistic self-consistent mean-field framework for atomic nuclei. *Comput. Phys. Commun.* **2014**, *185*, 1808–1821. [CrossRef]
50. Zhao, Q.; Ren, Z.X.; Zhao, P.W.; Meng, J. Covariant density functional theory with localized exchange terms. *Phys. Rev. C* **2022**, *106*, 034315. [CrossRef]
51. Heyde, K.; Wood, J.L. Shape coexistence in atomic nuclei. *Rev. Mod. Phys.* **2011**, *83*, 1467–1521. [CrossRef]
52. Quan, S.; Liu, W.P.; Li, Z.P.; Smith, M.S. Microscopic core-quasiparticle coupling model for spectroscopy of odd-mass nuclei. *Phys. Rev. C* **2017**, *96*, 054309. [CrossRef]
53. Pan, C.; Cheoun, M.K.; Choi, Y.B.; Dong, J.M.; Du, X.K.; Fan, X.H.; Gao, W.; Geng, L.S.; Ha, E.; He, X.T.; et al. Deformed relativistic Hartree-Bogoliubov theory in continuum with a point-coupling functional. II. Examples of odd Nd isotopes. *Phys. Rev. C* **2022**, *106*, 014316. [CrossRef]
54. Li, Y.; Zhou, E.F.; Yao, J.M. Benchmarking rotational correction energies in odd-mass nuclei. *arXiv* **2024**, arXiv:2409.12402.

MDPI AG
Grosspeteranlage 5
4052 Basel
Switzerland
Tel.: +41 61 683 77 34

Particles Editorial Office
E-mail: particles@mdpi.com
www.mdpi.com/journal/particles

www.ingramcontent.com/pod-product-compliance
Lightning Source LLC
Chambersburg PA
CBHW041340210926
51850CB00013B/2938
* 9 7 8 3 7 2 5 8 4 4 1 1 1 *